The Architecture of Space-Time in the Novels of Jane Austen

Ruta Baublyté Kaufmann

The Architecture of Space-Time in the Novels of Jane Austen

palgrave
macmillan

Ruta Baublyté Kaufmann
Haute Ecole Arc Ingénierie
Neuchâtel, Switzerland

ISBN 978-3-030-07914-7 ISBN 978-3-319-90011-7 (eBook)
https://doi.org/10.1007/978-3-319-90011-7

Softcover re-print of the Hardcover 1st edition 2018

Cover illustration: Escaflowne - iStock / Getty Images

Printed on acid-free paper

This Palgrave Macmillan imprint is published by the registered company Springer International Publishing AG part of Springer Nature.
The registered company address is: Gewerbestrasse 11, 6330 Cham, Switzerland

To Pierre, Simas and Mantas

Acknowledgements

I would first like to thank Margaret Tudeau-Clayton, for her guidance, patience, and continuous support. I would also like to thank John Mullan and Garrett Stewart, for their insightful comments and suggestions. In addition, I am grateful to Nancy Armstrong and Stephanie Trigg for their encouragement at early stages of my work, as well as to Egle Kackuté for giving me a friendly push to start it. Finally, I would like to thank my family and friends for their love and support.

Contents

1 Introduction 1

2 Changing Seasons: The Cyclical and the Linear 19

3 Changing Homes: Transformations in Space and Time 81

4 Spatiotemporal Movement and Psychological Change: Walking and Dancing 133

5 Conclusion 173

Index 177

Abbreviations[1]

E	*Emma*
MP	*Mansfield Park*
NA	*Northanger Abbey*
P	*Persuasion*
P&P	*Pride & Prejudice*
S	*Sanditon*
S&S	*Sense & Sensibility*
W	*The Watsons*

[1] Throughout I refer to the novels using these abbreviated forms of reference.

List of Tables

Table 2.1	*Northanger Abbey*	30
Table 2.2	*Sense and Sensibility*	30
Table 2.3	*Pride and Prejudice*	30
Table 2.4	*Mansfield Park*	31
Table 2.5	*Emma*	31
Table 2.6	*Persuasion*	32

CHAPTER 1

Introduction

Anyone who has heard of Jane Austen is likely to be aware of the "universally acknowledged" "truth" that "a single man in possession of a good fortune, must be in want of a wife" (*P&P* 3). After having been written by Austen, the phrase has been quoted over the last two hundred years by her readers and by literary critics, uttered on the screen by numerous Elizabeth Bennets or back-screen voices, as well as recycled in innumerable sequels and spin-offs of Austen's most famous novel. Countless other phrases, characters and situations have similarly been repeatedly re-read, re-thought, and re-embodied, for instance, in films, so that, as R. G. Dryden and L. Raw put it, Austen "has become a global traveller" (1). The rereading of Austen's work seems, moreover, to have taken a multitude of directions. Commenting on Austen's myriad "afterlives," Claudia L. Johnson summarizes:

> "Jane Austen" has mobilized powerful and contradictory ideas and feelings about taste, history, class, nationality, desire, manners, intimacy, language, and the everyday for very diverse readerships, and their complex histories of emotional, intellectual, and imaginative faith in Austen can inspire us to reread Austen in surprising and stunning ways. (15)

Many scholars and writers have considered the perennial appeal of the novels trying to determine the reason, or the reasons, that impel us to plunge into their worlds again and again. According to Harold Bloom, for

R. Baublytė Kaufmann, *The Architecture of Space-Time in the Novels of Jane Austen*, https://doi.org/10.1007/978-3-319-90011-7_1

example, Austen "created personality, character, and cognition; she brought into being new modes of consciousness" thus, inventing us and explaining us (v). Susannah Carson adds that "Austen novels haunt us our entire lives, inform our understanding of what it is to be human, and in the end fuse so wholly with our thoughts and feelings that it would be difficult to imagine the sorts of people we might have become had we never encountered them" (xii). For Eudora Welty, it is "the exuberance of her youthfull characters" and the novels' "sheer velocity" that "seems to vibrate through time" (9). Most recently, Devoney Looser in her book *The Making of Jane Austen* (2017) looks at the history of Austen's influences and her reception by analysing "particularly important stories of her making as they unfolded in culturally significant zones" (2). She points out that being referenced "by almost every category of people [...] Austen comes to us now as a household name" (2, 4).

As a passionate re-reader of the novels, I was curious to answer this question for myself: why has reading Jane Austen become a cyclical activity for me as it has for millions of other Austen lovers? I realized that even though I appreciate the work of the many critics who have placed the novels in their historical context I am drawn to what is not time bound, thus, what is "beyond" the context. The currently dominant historicist approach does not explain the perennial appeal of Austen's novels, which are not valued by most readers merely as historical documents. I decided to turn away from historicism and to take a look rather at the inner architecture of the texts.

In my work I want to argue that there are recurrent spatiotemporal patterns and structures in all six novels that constitute a source of enduring, if unconscious pleasure. In what follows I point out the overlapping natural and cultural cycles in the novels that co-exist in a constantly transmuting space-time and that are counterpointed with the linearity of pivotal events that drive the plot forwards. I examine the psychological relations of the characters, principally the heroines, to these space/time patterns, especially the transformations of their emotional states that prompt linear leaps. With regard to characters, I thus take the position of John Wiltshire, who treats them "as psychological realities" asserting that "imagining characters as actual beings is the primary, natural act of reading the realist novel, and the re-reading is a poor thing if this is lost sight of" (Wiltshire 2014, ix).

My project is then a formal as well as psychological analysis. It requires a level of generality since it seeks to identify, compare, and contrast patterns across the novels. Hence, though there are examples from the novels throughout, there is not a great deal of close reading except where this directly serves to illustrate the argument. In other words, this project requires reading at a certain distance.

Cyclical and Linear Space-Time

I am interested in looking at space and time as a four-dimensional concept. This concept is elaborated in Doreen Massey's book *Space, Place & Gender* (1994).[1] Responding to Foucault's and Jameson's arguments about a contrast between temporal movement and a notion of space, she claims that "the temporal movement is also spatial; the moving elements have spatial relations to one another" (264). She insists that we should not think of a linear process which reduces space to 3D (or even 2D), but accept four-dimensionality, for "space is not static, nor time spaceless" (264). She argues that there is a need to stop formulating space/time in terms of a dichotomy and points rather to Simone de Beauvoir's distinction between cyclicity,[2] gendered female, and "real change," gendered male. Cyclicity is central to immanence and "real change" to transcendence. Beauvoir's position was that females should take hold of the transcendent. Massey thinks that the two dichotomies (immanence/transcendence and space/time) are related. Moreover, according to her, Einstein's theory of four-dimensional space-time does not imply that difference between spatial and temporal is non-existent. Rather, she argues that "space and time are inextricably interwoven" (261).

My goal is to analyse space in relation with time in the novels of Jane Austen by taking into account its cyclical and linear aspects. I understand the concepts of cyclical and linear as developed by Jean Baudrillard in *The Illusion of the End* (1994). Linear, and spatialized, time for him is time that

[1] M. Bakhtin had already worked with the concept of space/time. According to E. Vlasov, he started the discussion and introduced the term of "chronotope" in 1920–1930 (Vlasov 1995, 39). Vlasov explains that Bakhtin "borrows the basic term chronotope from the natural sciences and applies it to literary theory to express the inseparability between three-dimensional space and time as the fourth dimension of space." Bakhtin defines the chronotope as "the intrinsic connectedness of temporal and spatial relationships that are artistically expressed in literature" and understands it as "a formally constitutive category of literature" (Bakhtin 1981, 84)" (Vlasov, 42).

[2] The term "cyclicity" is still absent from the *OED*, but it is included in several other dictionaries, which authorize my use of it. Merriam-Webster, for example, defines it as "state of being cyclic"; Collins explains it as "the quality or nature of being cyclic."

has a sense of direction; it is the "time of no return" (Clarke and Doel 31). On the other hand, cyclical time is the "time of eternal return" (Clarke and Doel 31), it is manifold. On the same principle, space can also be regarded as linear or cyclical. Linear space is represented as the straight line, marking "the shortest path between two points" (Baudrillard 10), while cyclical space with its sphericity "deflects all trajectories." To illustrate this concept, I would like to make a brief reference to the image of Bath in *Persuasion* (*P*) and *Northanger Abbey* (*NA*). Anne Elliot reluctantly goes to Bath as her new home—she cannot stay in her childhood house. Thus, Bath is her "final" destination, at least while she is unmarried, and represents a linear space. In *NA*, however, Catherine Morland goes to Bath to enjoy the season and the multitude of possibilities that this town can offer. In this case, it functions as a cyclical space open to various opportunities and trajectories.

The first phenomenon that comes to mind when reading Austen's novels from the angle of cyclicity and linearity is the changing seasons, which is the focus of Chap. 2. The cycle of nature is metonymically linked to corresponding agricultural activities which, together with weather, determine peoples' everyday life: what they will have on their table, their mobility; their activities; where they will stay—in other words, what spaces they will occupy and how they will spend and experience their time. The season modifies space/time in a cyclical, ever-repetitive way. For example, the female characters of *Pride and Prejudice* (*P&P*) survive through long months of Winter in the countryside wrapped in their warm shawls and lassitude, while others, more fortunate, dance through the animated London season, as in *Sense and Sensibility* (*S&S*). Both ways of experiencing Winter are repeated every year.

Next to the seasonal and agricultural cycles, there are the liturgical and calendar cycles, which are intertwined with the cycle of human life in general, and with the social cycles of everyday life. Thus, I show how the rhythms of these natural and cultural cycles coexist. For example, Autumn with its harvest and decaying nature corresponds to the new beginnings in physical, social, and emotional senses in the novels. Autumn is when the Miss Dashwoods move to Devonshire to start their new, socially different life after having lost their childhood home. But it is also in Autumn that both heroines meet the men they fall in love with and/or marry later on. Similar patterns are distinguishable in all six novels. Moreover, there are pivotal moments in every novel that approximately correspond to Christmas, Easter, Midsummer, and Michaelmas, thus emphasizing the importance of the liturgical and, possibly, the old pagan calendar to which the liturgical calendar is related. Within this likeness there are differences between the novels which I discuss using a table that charts the presence (or absence), duration, and importance of each season in the individual novels. This allows me to consider how Austen's use of these patterns evolves, or simply changes, over the course of her writing career.

During the time of the novels, then, while they follow the cycle of the year with its "usual" events, the heroines live their crucial year which will bring them to the totally new situation of their lives. For example, in *P&P* Elizabeth meets Mr. Darcy in a public ball in Meryton. These kinds of balls take place at regular intervals and are destined to amuse the young people of the neighbourhood and help them find suitable partners. However, this time, and from this point on, Elizabeth's year will not resemble her preceding seasons and at the end of this year she will find herself the mistress of Pemberley—far away from her childhood home in geographical, social, and psychological terms. Thus, the linear aspect of moving towards the space/time of no return coexists with the cyclicity of seasons and social rituals. The cyclical and the linear and their interconnection also relate to the inner lives of the heroines and their movement in space, as I will show.

Of relevance to my analysis is the work on cyclical and linear time by the founder of cultural semiotics Yuri Lotman. In his book *Universe of the Mind* (1990), he uses these terms when analysing the semiosphere in relation to the problem of plot. He describes as cyclical something that happens over and over, something unchangeable. He defines it as a principle. The linear, in that case, represents an event, an anomaly in opposition to regularity. Lotman points out that we often witness the interaction and mutual interference of the cyclical and the linear in literary texts. Writing about myths in archaic societies Lotman points out how the story reflects the cycles of nature, marked by the absence of a beginning and end: "Human life is looked at not as a line running from birth to death, but as an endlessly repeated cycle. The story, then, can begin at any point, and that point will serve as a beginning for the narrative which itself is a partial manifestation of the Text without beginning or end" (152). Looked at from this perspective, the novels of Jane Austen seem to contain a strongly integrated feeling of the continued cyclicity of seasons and of human life. However, this cyclicity is disturbed by unexpected events even if these events have been often hoped for by the characters: for example, the arrival of a young rich gentleman as a new neighbour breaks the cycle of disappointing balls and promises a real change. Such events knock out of balance the "natural" social routine, however reassuring it may be, and create psychological tension in the principal households. The reader meets these households in a transitional space/time, where different trajectories are still possible, but where one irretrievable direction will have to be taken, like the event of modernity, which irrupts into the cyclical life of what was still a dominantly agricultural society, as Tony Tanner points out.[3]

In Chap. 3, I analyse the movements that transform space/time. I examine how by following specific spatial and temporal trajectories the characters open up what is called in psychogenealogy a topography of the subconscious.[4] From this point of view, it is important not to consider the places they inhabit as absolute life spaces, but to examine the connections between different places as well as the possibilities to pass from one place to another, to leave or to appropriate them. These passages can be marked by different kinds of thresholds and by connecting spaces like windows, doors, or letters. The trajectories that characters take when moving through space/time are also marked by transitional space/time in, or through which the main psychological events take place.

Some of the characters, like Marianne Dashwood, repeat the ancestral cycle by remaining in the same social sphere, while others, like Elizabeth Bennet, push into newly and definitely transformed space. How do they experience this transformation and what are the reasons that lead them to it? What I see as a potentially problematic point, which often appears at the very beginning of the novels, is that the so called "ancestral cycle" in most cases belongs to the past: the present situation is messy, complicated, unbalanced. We see households such as Norland, Kellynch Hall, Hartfield, or Northanger Abbey not functioning as they are implied to have done in the past. They represent, as Lucy Worsley explains in her book *Jane Austen at Home* (2017) "the homes loved, lost, lusted after" (2). So, the trajectories that their occupants can or (as in the case of Emma and Eleanor Tilney) cannot take are conditioned by the present states of

[3] Tony Tanner in his book *Jane Austen* explains Austen's attachment to Tory values of the traditional rural lifestyle and her awareness of the new perception of the land—"speculative, acquisitive, calculating and irrelevant," together with "a whole range of new energies and impulses, new ideas and powers, which were changing or about to change England" (145). He adds that by *Persuasion,* Austen has changed her views and has started situating English society "in between an old social order in a state of decline and desuetude, and some new 'modern' society of as yet uncertain values, hierarchies and principles" (249).

[4] See Ulivucci, 33–59.

the households. I see in these situations a projected linearity: characters are leaving with the hope of returning, or staying with the hope of being able to leave one day. Other characters are able to break the cyclicity because of their inner motives. For example, Elizabeth Bennet wishes to move out of her parental house even though she has the opportunity of restarting the cycle of Longbourn by marrying Mr. Collins.

It is at this point that I draw on Gaston Bachelard's *The Poetics of Space* (1958) and his phenomenology of houses in relation to the universe that surrounds them. First, I examine characters' relationships with their childhood home as the "original shell," then, look at the intimate spaces that characters recreate in the houses they visit, thus constructing a shelter from the hostility of the "universe." The secure space unites past, present, and future and, in that way, forms a cycle. For example, in *Pride & Prejudice* such a space is Longbourn: it seems to be crucial for Mrs. Bennet to save Longbourn for her own children. And as her daughters get married and have to move, she wishes them to stay as close to "the nest" as possible. The intimate spaces are also important because they host the "introspective moments" experienced by the heroines of the novels, which propel the situations to their resolution. These are important internal events after which nothing is as it was before. One such moment is triggered by Mr. Darcy's marriage proposal during Elizabeth Bennet's visit to Charlotte Collins in Hunsford. In all the novels, visiting people and staying with them for several weeks is part of a usual, ritualized way of life as is the case with this visit, which seems to be a common cyclical experience. However, Mr. Darcy's marriage proposal brings a change and Elizabeth plunges into a transitional space-time of introspection, which will lead her to a different self-understanding.

Bachelard sets the childhood home in opposition to the dream home. In Jane Austen, the dream home *par excellence* is Pemberley. For Elizabeth Bennet it contains all the perfect qualities she could wish a home to have and which are all that her childhood home—Longbourn—lack. The recognition of the dream home helps her make that linear change: a move onto a superior level in social, economic, and psychological senses. In this respect, I see a parallel between Elizabeth and her mother, who, being a daughter of an attorney, leapt into her own dream house, Longbourn, by marrying a gentleman.

The fourth type of home, that I distinguish in the novels, is an independent home, created by conscious choice—an adult home. It marks the final stage of the spatiotemporal trajectory. In some cases, like Emma

Woodhouse's, it is the same house, the same home space that changes its emotional and social status as linear events evolve in time. However, not all Austen's characters have a fixed, permanent home. This privilege is denied to single adult females, like Caroline Bingley or Mary Crawford, who depend on their brothers and possibly never found their own household.

In Chap. 4 I concentrate on spatiotemporal movement: walking and dancing. Like space and time, movement can be cyclical and linear. This is also the case of the body—the initiator of movement. In *The Production of Space* (1991) Henri Lefebvre states that "the body unites cyclical and linear, combining the cycles of time, need and desire with the linearities of gesture, perambulation, prehension and the manipulation of things" (203). This combination unfolds in dancing and walking which occupy much space/time in the novels. In dancing there is the importance of rhythm and the organized gestural system. It is also an emotionally charged, face-to-face experience which changes the individual perception of time and space. How do the moving bodies move space-time? How do the characters live this cyclical and linear experience? Similar questions can be asked concerning walking, which differs from dancing by being rather a solitary or group act. Though a repeated and ritualized activity, walking can also host linear events, such as Mr. Darcy's giving the explanatory letter (which reveals the truth of his relations to Mr. Wickham) to Elizabeth during her solitary walk in Kent.

Historically, walking was the main physical activity for genteel women in Austen's time.[5] However, next to this physical aspect, walking allows a moment of privacy to think or to share intimate "secrets" with a chosen partner. Moreover, in most cases, walking helps to deal with agitated emotions and to find a more harmonious psychological state. Dancing, on the other hand, often generates strong and contradictory emotions, as I show in Chap. 4, thus making it a more problematic and tense experience than walking.

In general, dancing is first of all an experience of body in space/time. The rhythm of the music meets the rhythm of the steps and of the spatial trajectory. The physical closeness changes the perception of space/time by internalizing it and making it completely dependent on the emotional exchange between partners as we can see in the succession of uncomfortable silent moments and sharp, destabilizing remarks between Elizabeth and Mr. Darcy when they dance in Netherfield. Externally the dancers are led

[5] See Todd, 337.

by the cyclicity of the choreography while internally they might live a linear experience. The ritualized character of the actions of the body, timings, and configurations, as well as the physiological rhythm of steps, breath, and, possibly, talking is also a choreography that encloses the linearity of psychologically intense moments such as falling in love, forming an opinion and making a radical decision, "knowing oneself" or exchanging some important information that will radically modify the course of events.

Finally, we can also look at the reception of the novels of Jane Austen from this angle of space/time. The numerous sequels that have been written evoke cyclicity. Situations, plot lines, the essential questions touched in the novels do not seem to be resolved, they are kept in suspension, incomplete, and out of time. Readers refuse to accept the actual beginning and the end of the novels, as well as the chosen points of view. There are for instance sequels retelling *Pride and Prejudice* from Mr. Darcy's, or Mary Bennet's point of view. Indeed, D. S. Lynch argues that Austen herself prepared the ground of the sequels of her novels by almost "scripting the terms of her future sequelisation" (167). One of the examples she gives is that of Fanny Price's sister Susan introduced into *Mansfield Park*, which allows "a cycle to recommence" (167).

Space and Time in Austen: A Review of Critical Work

Research on space and time in Jane Austen's work has frequently been conducted from the perspective of social history and geographical exactitude. This is, for example, the case in Deidre Le Faye's book *Jane Austen. The World of Her Novels* (2002) where the author gives a detailed analysis of topography, architecture, and typical interiors of the time. She also provides a detailed description of social, cultural, and family life at the end of the eighteenth century and the beginning of the nineteenth, taking into account biographical facts about Austen and drawing on her letters. She names the possible places behind the houses and locations in all six novels and reproduces portraits of people who might have resembled the main characters. In short, Le Faye is trying to ground the novels in the historical reality of the time.

A similar purpose is implied in the even more detailed collection, edited by Janet Todd, *Jane Austen in Context* (2005). Alongside thorough analyses of historical and cultural contexts, including chapters on agriculture, cities, domestic architecture, pastimes, or psychology, as well as biographical facts, we find a brief survey of critical responses to the novels. Daniel

Pool in *What Jane Austen Ate and Charles Dickens Knew* (1993) similarly offers an extensive description of nineteenth-century everyday life, explaining its rules, customs, and habits. More specifically, Simon Varey's historicist study *Space and the Eighteenth-Century English Novel* (1990) deals with architectural conceptions of space taking into account their political and theological implications. Varey discusses the "architectural imagery" in the novels of Defoe, Fielding, and Richardson declaring that "spaces created (in theory or practice) by architects and those created by the novelists [...] express specific ideologies and are therefore political" (4). Though these detailed historical studies are useful, my goal, as I indicated earlier, is to turn from this approach to examine a more atemporal aspect of Austen's work—its recurring cyclical and linear structures.

More pertinently John Mullan, through the prism of historically grounded specific questions such as "How much does age matter?," "Why is the weather important?," or "Why is it risky to go to the seaside?," reaches into the internal world of the novels, seeking, as he says himself "to catch her [Austen] in the act of greatness, by scrutinising the patterns and puzzles that she builds into her novels" (Mullan 2012, 5). Thus, as well as explaining some of the historical issues which Austen "would have taken for granted" (7) he gives us insights into the psychological and structural complexity of her novels. This shows that he is also interested in the perennial appeal of the novels, which he finds beyond the historical specificities of context, in patterns and puzzles. However, he does not deal with spatiotemporal patterns and rhythms as I do. My aim is different too, since I look at cycles and linear events in relation to the psychological trajectories of the characters.

Work on time often deals either with the chronology of events, as in Ralph Nash's "The Time Scheme for Pride and Prejudice" (1967) or with the historical moment of the end of the eighteenth century and correspondences between historical moments, as in Laura Caroll's "A Consideration of Times and Seasons" (2003). Nevertheless, the detailed time-lines, fully and precisely elaborated by D. M. Shapard in his annotated editions of the novels, have been extremely useful throughout my work, especially in Chap. 2.

Some research, which I will now consider, adopts a more philosophical approach, mostly by examining various aspects of private and public spaces. The following scholars refer to the work of G. Bachelard, as I do in my sections on childhood and dream homes, but they do not look at space/time in relation to the heroines' psychological trajectories through different types of homes. In *The Spaces of Privacy in Nineteenth Century Fiction* (1975) Frances Hart looks at the issue of privacy from a historical point of view, explaining the new desire for privacy which appears in the eighteenth century. Drawing on Bachelard's *The Poetics of Space* and his analysis of intimate space as well as miniaturization she argues "as a character grows, matures, and accumulates painful or distressing sympathies, secrets, or feelings of shame for mistakes made, the character discovers the need for privacy" (319).

In "Exploring Space: the Constellations of *Mansfield Park*" (1992) John Skinner goes through different types of space in the novel: public or social space as expressed through the external connections of the Great House, interpreted in terms of "conventional topography"; private or intimate space, where, like Hart, he uses Bachelard's phenomenological approach to analyse the house in connection to the subjective experience of Fanny; existential space represented by the other young people; the novel as diegetic space (referring to sites of ambiguity in the text); and hermeneutic space produced by all these kinds of space. At this point he turns to intertextuality using Bruno Bettelheim's psychoanalytical study of fairy tale motifs and examines the parallels between *Mansfield Park* and Cinderella.

In his book *Jane Austen* Tony Tanner also dedicates a couple of pages in his chapter about *Northanger Abbey* to the psychological processes of Catherine when confronted with the "mysterious" wardrobes, drawers, and chests of her bedroom in the Abbey. He refers to the chapter "Drawers, Chests and Wardrobes" in Bachelard's *The Poetics of Space* and presents the potential psychological reading of these intimate spaces. Meanwhile Freya Johnston in her article "Public and Private Space in Jane Austen" (1997) looks at the "reflective interiors" not only in the novels, but also in Austen's biographical circumstances. She claims that there are points of intersection between Austen's fictional work and her letters and thus, looks for correspondences between these two types of spaces.

In "Prospect and Refuge in the Landscape of Jane Austen" (2001) Barbara Britton Wenner also draws on Bachelard to discuss *Persuasion.* In addition, she looks at the function of landscape from a historical and gender perspective. She argues that Jane Austen's heroines have a way of seeing and being in the landscape, without adopting either the view of the male as the landscape proprietor or the attitude of the passive female who "becomes one with nature." They look for refuge and prospects that help them find their "felicitous space." Bachelard's images of refuge in relation to the heroines' personal identity are also evoked in Laura Mooneyham White's "Traveling to the Self: Comic and Spatial Openness in Jane Austen's Novels" (1998). Leland Person in "Playing House: Jane Austen's Fabulous Space" (1980) compares the retreat of heroines to spaces of refuge with Bachelard's "hut dreams." Raymond Hilliard, on the other hand, takes a rather different view of private space and the search for refuge in "Emma: Dancing with no space to turn in" (1973). According to him the characters of *Emma* suffer from confinement and wish to enlarge the circle. He also comments on how they constantly live in future time. Tony Tanner seconds his thought adding that in "Emma's society there is no room for manoeuvre, no room for rearrangement, no room for any kind of escape" (190).

Julie Park in "What the Eye Cannot See: Interior Landscapes in *Mansfield Park*" (2013) also examines the relationship between the external space and internal space/sight of the heroine. She is interested in the moment "that marks the boundary between external reality and internal reality, the realm of the imagination" and which transforms visual perception "into internal sight" (170). She gives several examples of the coexistence of these external and internal landscapes in *Mansfield Park,* as in the passage on Sotherton where the reader experiences "the visible landscape itself that the characters feel beneath their feet as they walk through it, and the invisible landscape of Fanny's mind with its own turns and episodes that hold off the awareness of being excluded" (173).

One of the first studies on space in Austen's novels was Alistair M. Duckworth's book *The Improvement of the Estate* (1971). The starting point of his argument is the fact that all Austen's heroines are haunted by the potential danger of ending up like Mrs. Smith in *Persuasion.* According to him "though she appears at the end of Jane Austen's writing life, Mrs Smith has always existed as a latent possibility in the novelist's thought, an unvoiced threat, the other possible pole of existence" (3). This danger is perceived by the characters not only on a social level, but also on a metaphysical and theological level, for being deprived of "stable and inherited estate" means being evicted from the "grounds of being and action" (4). However, he notices that the heroines respond in a very positive way to this "insecure world" with their "individual energy" (7). Duckworth leads us through the novels' estates and landscapes, which he sees as reflections of their owners' moral and social character. He also analyses the historical trend to improve and innovate these traditional spaces, symbols of the "inherited culture," (55) and the different ways this is done. He distinguishes between those who "improve" by neglecting the traditional and the functional and displaying their individualism, like the subversive Crawfords in *Mansfield Park* (54), and those who continue in the "natural" vein with the respect to the past, like the responsible Mr. Knightley in *Emma* (156).

More pertinent to my approach is Julianne Pidduck's article "Of windows and country walks: frames of space and movement in 1990s Austen adaptations" (1998). Pidduck takes a topographical and gender perspective to investigate gendered space through the specificity of female interiors, the female gaze through the window and the "social choreography of movement" in cinematic adaptions of the novels. She remarks that "in contrast to the precise dialogue and intricate human interaction condensed into Austen's parlours, libraries and balls, exterior sequences [...] tend to create a sense of spatial and emotional expansiveness, not unlike the role of dance numbers in the musical" (382). I take up these same features of Jane Austen's novels in Chaps. 2 and 3, though I look rather at the cyclical and linear aspects of walking and dancing, taking into account their physical, emotional, and social implications.

In "'Young ladies are delicate plants': Jane Austen and Greenhouse Romanticism" (2010), Deidre Shauna Lynch develops an interesting argument that also bears on my thesis. Her goal is "to explore what happens when greenhouse romanticism meets the history of novelistic realism" (692). She uses the term *greenhouse romanticism*, as opposed to *green romanticism*, to underline the "romantic-period culture's favourite scientific instrument for both apprehending and augmenting Nature's plurality" (692). Drawing a parallel with exotic flowers grown in greenhouses, she situates the physical and psychological development of Austen's heroines, especially Fanny Price and Catherine Morland, in "artificial climates" and claims they are boosted by "artificial heat" (690, 718). Lynch concludes that "exploiting the plurality and pliability of romantic Natures, Austen aligns the second nature of education with the florist's flowers of the greenhouse and through that alignment rewrites that plot of naturalization in an idiom of artifice" (719). I take a rather different perspective: without opposing nature and artifice, which is a historically grounded opposition, thus, one that is not pertinent to the question of the perennial appeal of the novels, I examine the superposition and coexistence of natural and cultural cycles in the novels and their relation to the inner development of the characters.

To return to the treatment of time in the novels, Loraine Fletcher in "Time and mourning in *Persuasion*" (1998) looks at the time of loss, the rituals of mourning, the pressure of passing time on the heroine, and the opening up to the transcendent, all in the historical context. Similarly, though more pertinent to my thesis, William Deresiewicz in *Jane Austen and the Romantic Poets* (2004) explores *Persuasion* through the phenomena of memory and mourning. He goes further than Fletcher, claiming not only that Anne's mourning "violates nature" (134), but that the whole Elliot family is presented through "the theme of sterility" and "the refusal of natural forms of continuity" (135) from which Anne has to liberate herself. In general, Deresiewicz perceives the relationship between time and events of the three later novels as salt in water, that is, he finds that "it scarcely makes sense even to distinguish between 'events' and 'time'" (43). However, he sees much less structural complexity in the earlier novels arguing that "time here [in *NA*, *S&S*, and *P&P*] is essentially Bergsonian *temps*—linear, unidirectional clock-time—with none of the thickening and deepening provided by those back-and-forth movements of consciousness that cause time present and times past to coexist in a complex temporal space" (37). In Chap. 2, I hope to demonstrate that Austen's early novels also contain complex spatiotemporal structures and that linearity is not only disruptive, but also intertwined with cyclicity.

Thus, the originality of my work lies in its uncovering of the spatiotemporal architecture of each and all of Austen's novels, and the ways that the linear and cyclical are intertwined in patterns that are variously connected to the psychological trajectories of the principal characters.

Works Cited

Primary Sources

Austen, Jane. *Northanger Abbey*. Ed. Barbara M. Benedict and Deidre Le Faye. Cambridge: Cambridge University Press, 2006a.

Austen, Jane. *Sense and Sensibility*. Ed. Edward Copeland. Cambridge: Cambridge University Press, 2006b.

Austen, Jane. *Pride and Prejudice*. Ed. Pat Rogers. Cambridge: Cambridge University Press, 2006c.

Austen, Jane. *Mansfield Park*. Ed. John Wiltshire. Cambridge: Cambridge University Press, 2005a.

Austen, Jane. *Emma*. Ed. Richard Cronin and Dorothy McMillan. Cambridge: Cambridge University Press, 2005b.

Austen, Jane. *Persuasion*. Ed. Janet Todd and Antje Blank. Cambridge: Cambridge University Press, 2006d.

Secondary Sources

Bachelard, Gaston. *The Poetics of Space*. Boston: Beacon Press, 1994.

Baudrillard, Jean. *The Illusion of the End*. Trans. Chris Turner. Cambridge: Polity Press, 1994.

Bloom, Harold. "Preface" in *A Truth Universally Acknowledged. 33 Great Writers on why we read Jane Austen*. Ed. S. Carson. New York: Random House, 2009. v–vi.

Caroll, Laura. "A Consideration of Times and Seasons." *Literature/Film Quaterly* 31.3 (2003): 169–176.

Carson, Susannah. "Introduction" in *A Truth Universally Acknowledged. 33 Great Writers on why we read Jane Austen*. Ed. S. Carson. New York: Random House, 2009. xi–xx.

Clarke, David B. and Doel, Marcus A. "Jean Baudrillard." *Key Thinkers on Space and Place*. Eds. Phil Hubbard, Rob Kitchin and Gill Valentine. London, Thousand Oaks and New Delhi: Sage Publications Ltd, 2004.

Deresiewicz, William. *Jane Austen and the Romantic Poets*. New York: Columbia University Press, 2004.

Duckworth, Alistair M. *The Improvement of the Estate*. London: Johns Hopkins, 1971.

Fletcher, Loraine. "Time and Mourning in Persuasion." *Women's Writing* 5.1 (1998): 81–90.

Hart, Frances. "The Spaces of Privacy: Jane Austen." *Nineteenth-Century Fiction* 30.3 (1975): 335–350.

Johnson, Claudia L. *Jane Austen's Cults and Cultures*. Chicago: University of Chicago Press, 2014.

Johnston, Freya. "Public and Private Space in Jane Austen." *English: The Journal of the English Association* 46 (1997): 193–212.

Lamont, Claire. "Domestic architecture." *Jane Austen in context*. Ed. Janet Todd. Cambridge: Cambridge University Press, 2005. 225–233.

Le Faye, Deirdre. *Jane Austen. The World of her novels*. London: Frances Lincoln, 2002.

Lefebvre, Henri. *The Production of Space*. Trans. Donald Nicholson-Smith. Oxford: Blackwell, 1991.

Lynch, Deidre Shauna. "Sequels." *Jane Austen in context*. Ed. Janet Todd. Cambridge: Cambridge University Press, 2005. 160–168.

Lynch, Deidre Shauna. "'Young ladies are delicate plants': Jane Austen and Greenhouse Romanticism." *English Literary History* 77.3 (2010): 689–729.

Looser, Devoney. *The Making of Jane Austen*. Baltimore: Johns Hopkins University Press, 2017.

Lotman, Yuri M. *Universe of the Mind*. Trans. Ann Shukman. Bloomington and Indianapolis: Indiana University Press, 1990.

Massey, Doreen. *Space, Place and Gender*. Cambridge: Polity Press, 1994.

Mullan, John. *What Matters in Jane Austen?* London: Bloomsbury, 2012.

Nash, Ralph. "The Time Scheme for Pride and Prejudice." *English Language Notes* 4 (1967): 194–198.

Park, Julie. "What the Eye Cannot See: Interior Landscapes in Mansfield Park." *The Eighteenth Century* 54. 2 (2013): 169–181.

Person, Leland. "Playing House: Jane Austen's Fabulous Space." *Philological Quarterly* 59.1 (1980): 62–75.

Pidduck, Julianne "Of windows and country walks: frames of space and movement in 1990s Austen adaptations." *Screen* 39.4 (1998): 381–400.

Pool, Daniel. *What Jane Austen Ate and Charles Dickens Knew*. New York: Touchstone, 1993.

Raw, Laurence and Robert G. Dryden, eds. Global Jane Austen. Pleasure, Passion and Possessiveness in the Jane Austen Community. New York: Palgrave Macmillan, 2013.

Skinner, John. "Exploring Space: The Constellations of Mansfield Park." *Eighteenth-Century Fiction* 4.2 (1992): 126–148.

Shapard, David M., ed. *The Annotated Emma. Jane Austen*. New York: Anchor Books, 2012.

Shapard, David M., ed. *The Annotated Northanger Abbey. Jane Austen*. New York: Anchor Books, 2013.

Shapard, David M., ed. *The Annotated Persuasion. Jane Austen*. New York: Anchor Books, 2010.

Shapard, David M., ed. *The Annotated Pride and Prejudice. Jane Austen*. New York: Anchor Books, 2007.

Shapard, David M. ed. *The Annotated Sense and Sensibility. Jane Austen*. New York: Anchor Books, 2011.

Tanner, Tony. *Jane Austen*. London: Macmillan, 1986.

Todd, Janet, ed. *Jane Austen in context*. Cambridge: Cambridge University Press, 2005.

Ulivucci, Christine. *Psychogénéalogie des lieux de vie, Ces lieux qui nous habitent*. Paris: Payot, 2008.

Varey, Simon. *Space and the Eigtheenth-Century English Novel*. Cambridge: Cambridge University Press, 1990.

Vlasov, Eduard. "The World According to Bakhtin: On the Description of Space and Spatial Forms in Mikhail Bakhtin's Works." *Canadian Slavonic Papers* 37.1–2 (1995): 37–58.

Welty, Eudora. "The Radiance of Jane Austen" in *A Truth Universally Acknowledged. 33 Great Writers on why we read Jane Austen*. Ed. S. Carson. New York: Random House, 2009. 9–17.

Wenner, Barbara. "Prospect and Refuge in Landscape of Jane Austen." *Journal for Eighteenth-Century Studies*, 33. 2 (2010): 265–266.

White, Laura Mooneyham. "Travelling to the Self: Comic and Spatial Openness in Jane Austen's Novels." *Critical essays on Jane Austen*. Ed. Laura Mooneyham White. New York: G.K. Hall; London: Prentice Hall International, 1998. 198–213.

Wiltshire, John. *The Hidden Jane Austen*. Cambridge: Cambridge University Press, 2014.

Worsley, Lucy. *Jane Austen at Home*. London: Hodder & Stoughton, 2017.

CHAPTER 2

Changing Seasons: The Cyclical and the Linear

All of Jane Austen's six novels integrate hitherto unrecognized aesthetic cycles. I distinguish several cyclical structures: the cycle of human life, the cycle of the changing seasons, the agricultural cycle, the liturgical and calendar cycle, the cycle of the social seasons, and the social cycle of everyday life. These different cycles are superposed or fitted into each other. The natural cycle of the changing seasons is the most prominent, together with the cycle of human life—birth, growth, death—to which all individuals are submitted. As I will show in this chapter, Austen's plots have an internal rhythm in which the changing seasons intertwine with the other cycles and overlap with linear events. For example, the agricultural cycle is shown to be closely linked to the seasonal cycle and, as farming still constituted the main source of income for the landed gentry,[1] its claims are placed above social ones. Thus, even a master of landed property, such as Mr. Bennet in *Pride and Prejudice* (*P&P*) can only complain about the situation that the horses "are wanted in the farm much oftener than [he] can get them" (*P&P* 34) and when these horses are actually engaged in the farm, his daughters Jane and Elizabeth have to ride or walk if they wish to visit Netherfield Park and its potential fiancé (*P&P* 35).

Austen arranges the events of her plots in relation to calendar and liturgical cycles: the houses are let at Michaelmas, and there are parties for

[1] See Todd, 185–192.

R. Baublytė Kaufmann, *The Architecture of Space-Time in the Novels of Jane Austen*, https://doi.org/10.1007/978-3-319-90011-7_2

Christmas and Midsummer. Notably, as the liturgical calendar follows the old pagan calendar and mirrors its feasts,[2] the key liturgical dates loosely correspond to the days of Summer and Winter solstices as well as to Spring and Autumn equinoxes, thus connecting the liturgical to the seasonal cycle. In the novels these periods—Christmas, Easter,[3] Midsummer (St. John for the liturgical calendar, but Austen uses the pagan term of Midsummer) and Michaelmas—often mark critical and pivotal moments in the narrative.

The social cycles are superposed on the seasonal and the liturgical cycles, and sometimes clash with the agricultural cycle. For the eighteenth- and nineteenth-century English gentry a year was divided into two social seasons: Summer in the country with its picnics, and Winter, preferably, in town (London or Bath in the novels), with its balls.[4] The Summer season comprehends the Summer months, Autumn with its hunting period, and even the beginning of Winter. Accordingly, the social Winter also includes Spring. The borders of these two seasons are not strictly defined: in the novels there is some fluctuation from one family to another. Thus, we see the Dashwood girls leaving for London with Mrs. Jennings right after Christmas, while the sportsmen, such as Sir John, delay their journey because of the fine weather for hunting (*S&S* 190).

In general, Jane Austen respects this traditional social schema in her novels. However, the presence and pressure of the agricultural seasons can be felt in her texts. Though the two cycles mainly function in parallel, from time to time they intertwine or even clash: as mentioned earlier, farming is still the main source of income for a large part of the landed gentry, and social needs must adapt themselves to agricultural imperatives. This is the case with Mary Crawford's harp in *Mansfield Park* (*MP*): she wants it delivered immediately, in order to fully display her accomplishments,

[2] According to Ronald Hutton, the pagan ritual calendar contains eight main points in the cycle of the year. These are the solstices, the equinoxes and four dates which marked the traditional beginnings of the seasons: the first days of May, August, November and February, which respectively started summer, autumn, winter and spring. All these days (or their eves) hosted important feasts such as May Day (on the first of May) and November Eve.

[3] I use Easter as a spring reference point because in Austen's novels this moment plays an important pivotal role. However, Easter was not one of the four Quarter Days on which rents and taxes were due and salaries were paid which were Lady Day (25 March) Midsummer, Michaelmas and Christmas. Before the Calendar Reform in 1752 (a switch from the Julian Calendar to the Gregorian) Lady Day was also the first day of the year.

[4] See Pool, 50.

but as it is the time of the hay-harvest no cart is available and Mary is obliged to wait (*MP* 68–69). She relates the incident to Mr. Bertram in a theatrical, exaggerated style: "guess my surprise, when I found that I had been asking the most unreasonable, most impossible thing in the world, had offended all the farmers, all the labourers, all the hay in the parish" (*MP* 68). Her ironic reaction to the delay, highlighted by the use of superlatives, exposes her blindness to the dependence of her social class on agricultural labour and so implies a judgement on her character.

Finally, this annual social cycle of the gentry together with the seasonal cycle enclose the social cycles of everyday life, composed of visits, trips, and lengthy stays, walks, balls, and dinners. These activities may take different patterns in different places or at different stages of life (for example, before and after marriage), but they are usually comparable to a ritual. In fact, these social cycles are dependent on all the others: the type and the frequency of activities are determined by the time of life of an individual, the natural season, the weather, the liturgical moment, and the social season. The weather is often the decisive factor in these cases and is closely observed, related, and talked of in the novels. It not only "clues us in to the passing of the year" (Mullan, 101), but also regulates the activities of everyday life. Therefore, it has a decisive impact on the novels' plots. As Mullan remarks Austen "makes her plots turn on weather" allowing "the weather to shape events" (101). Though there is, as he comments, an element of "chance" in this, the weather is also an expression of seasonal cyclicity.

My claims find support from D. S. Lynch who argues that, Austen constructed her narratives in a cyclical way "playing with the conventions of narrative teleology and closure" (Lynch 2005, 167). She remarks that her novels "often incorporate repetition—the recursive rhythms of everyday life—in ways that stay the forward momentum of the plot and qualify the readers' sense of 'hastening' pell-mell towards 'felicity'" (167). Moreover, she notices that Austen often arranges her novels "as though they were the sequels to earlier (untold) stories" (167). Thus, the reader is plunged into the narrative at the beginning of a cycle and, at the end of the narrative, the reader can foresee how the closing cycle is going to give birth to a new one. I extend and develop this argument, adding other types of cyclical structures (as mentioned earlier) present in the novels.

The main action of four of the novels evolves during the period of one year; the remaining two—*NA* and *P*—are shorter, although the idea of a year and changing seasons is still present: Catherine Morland, for example, marries Henry Tilney one year after their first meeting (*NA* 261). As mentioned before, each of the four seasons has its own defining aspects, its relations with the liturgical and agricultural calendars, and its specific social rituals. However, this cyclicity is frequently disrupted by linear events that make this particular year different from all the others. The reader follows one of the most important and transformative years in the lives of the novels' main characters. All the heroines go over the threshold that separates an unmarried girl from a married woman, which brings the major change in their social status. But before they get there, the female, as well as the male characters go through a psychological transformation which often closely mirrors the changing seasons.

The importance of acknowledging and accepting seasonal cyclicity can be illustrated by the scene of Marianne's farewell to Norland. For Rodney S. Edgecombe, a solitary wanderer, Marianne Dashwood mourns Norland's trees which she is about to leave, but her "ostensible purpose [...] to project the sublime indifference of the natural cycle to human affairs [...] falls short of that end" (609).[5] Her soliloquy shows her deliberate denial of natural processes as if they depended on her presence to observe them:

> And you, ye well-known trees!- but you will continue the same.—No leaf will decay because we are removed, nor any branch become motionless although we can observe you no longer! (*S&S* 32)

Marianne denies and wilfully ignores the seasonal cyclicity, recklessly throwing herself into the linear experiences of "no-return," which makes her the most dramatic of Jane Austen's heroines. Emma Woodhouse also tries to forcefully move forward the natural cyclicity of life. Fortunately, her endeavours tend not to impact the linear trajectory, thus avoiding traumatizing consequences. These examples indicate how much the influence of the changing seasons is present in Austen's novels. Representing spatio-temporal cyclicity the seasons condition and reflect different life stages on a social and psychological level and, at the same time, they determine everyday activities. Moreover, the novels seem to affirm the necessity to respect the seasonal cyclicity and follow it without trying to resist, inter-

[5] Edgecombe suggests that Austen's real intention here is to mock the Picturesque way of thinking: "for this tendency to immobilize, for this habit of thinking in tableaux, Marianne's cult of the Picturesque is partly to blame, since, in its popular form at least, that aesthetic is nothing if not the collation of freeze frames, freeze frames that the viewer privileges above such instances of dynamic evolution as the movement of a river, or the change of seasons" (611).

rupt, and/or wilfully change it. This is done by the characters insofar as they pace their walks according to the weather. On the other hand, linearity is always present, unavoidable, and often in fact wished and hoped for by the characters even if it is only to move on to a different cycle.

Autumn

Cyclicity and linearity are in close relationship in the novels. One of the examples of this phenomenon are the new beginnings that often occur in Autumn and which later will have crucial consequences and reveal themselves to have been linear experiences. In these instances, the reader senses the seasonal and the social cycles intertwined in the background, and, at the same time, follows the unexpected transforming events in the foreground.

In general, Autumn with its decaying nature and the new seeds sown in the ground brings new beginnings during the period of active social gatherings and new meetings. First, the month of September marks a limit between the importance of outside over inside notably in *S&S* and *P&P*, as I shall show later in the chapter. It is a month of crops for farmers and of bird shooting for genteel men. From October on, many social activities, especially for women, take place inside. Their dependence on the weather for opportunities to leave home becomes an issue. While men are mobile most of the time and we never read of them being in physical danger or provoking social judgement because of the unsuitable weather, women, being considered, as Mr. Woodhouse calls them, "delicate plants" (*E* 318) must not expose themselves to rain or dirt. Elizabeth Bennet neglects this convention when "crossing field after field at a quick pace, jumping over stiles and springing over puddles" (*P&P* 36) in order to visit her sick sister and she is consequently considered with contempt by the fashionable Bingley sisters. Similarly, Jane Fairfax is the object of unwelcome concern for her health in Spring because she has walked to the post office during the rain and might have got her feet wet, while Marianne Dashwood's wanderings in stormy weather in Autumn and Spring end up with more or less dramatic consequences for her health. This movement towards the inside evokes the cyclical time of Yuri Lotman's archaic culture in his book *Universe of the Mind* (151). It is repeated every year and is "synchronized with the cyclical processes of nature" (151). It is notable, however, that this cyclical synchronization applies to women more than men. Furthermore, what is going on inside cannot always be compared to hibernation for it is usually a busy time socially.

The Autumn months in all the novels (except *NA*) see social gatherings and new projects directed to promote the self-sufficiency of the social circles. This allows the country families to entertain each other when little novelty is possible. In *P&P*, *P*, and partly *S&S* the arrival of a rich and handsome bachelor promotes the revived social life of the neighbourhood. In *Emma* (*E*) and *MP* we also find young eligible bachelors at the centre of attention, although they have already been in the county for some time or were born there. It is the attraction of Mr. Bingley that causes Jane Bennet's (maternally pre-programmed) few days stay in Netherfield and that gets Elizabeth out of home to walk (an unforeseen) three miles "in all this dirt" (*P&P* 35) exposing "the brilliancy which exercise had given to her complexion" (*P&P* 36) to Mr. Darcy's admiration.

The movement towards the inside during this period can also be felt on the psychological level. During this time of intense social life some characters live the emergence and the development of very strong feelings, which they are not yet able to express outwardly or, if they do show them, these feelings cannot yet be legitimated. As suggested above we might compare this with seeds sown in the ground and lying there hidden full of potential to grow in the Spring, although, just as not all seeds will grow, not all newly born feelings will evolve into marriage. It is impossible to say when looking at these autumnal beginnings which of them will actually end in marriage at this stage of the plot. In *S&S* Marianne and Willoughby live their short and intense love story without any promise of marriage, while Elinor suffers from Edward's absence and ends up finding out about his prior engagement. The reader can observe Jane Bennet and Mr. Bingley who fall gradually more and more in love with each other, and Mr. Darcy who gets increasingly attracted to Elizabeth in *P&P*. Edmund and Mary Crawford are getting attached to each other, and Maria is developing a secret passion for Henry in *MP*. In *E* Harriet, under the influence of Emma, starts developing feelings for Mr. Elton, while Jane Fairfax and Frank meet in Weymouth and get secretly engaged. In *P* Anne and Wentworth meet again and their old love slowly begins to revive, while Louisa and Maria live their first love experiences.

Another aspect of Autumn in the novels is connected with the literal and figurative meanings of "fall" which, according to the *OED*, used to be a word for "autumn" in British English. When writing about Louisa's fall in *P* Deresiewicz remarks that "what goes up must come down, but by the same token, what goes down eventually comes back up—just as (the pun is inevitable) the season of 'spring' inevitably succeeds that of 'fall'" (140).[6] Louisa's fall is not the only autumnal linear fall—whether literal or figurative—in the novels. Marianne falls and it appears to be psychologically fatal to her; Jane "falls" ill at Netherfield and gets closer to Mr. Bingley; the theatricals in *MP* cause the "fall" of the outwardly perfect household. However, the linear quality of these falls starts a new cycle in a rearranged spatial and relational configuration and the Autumn in the novels contains the tension of the new hopeful beginning even if it is often closely followed by disappointments.

In general, then, during these autumnal months, Austen brings together and links the main characters and circumstances, necessary for the future development of the plot, which will stage the most significant year in the main characters' lives. The most important spatial changes take place at this time: the characters are forced to leave their original home (*S&S*, *P*, and *The Watsons* [*W*]), face a future departure (*P&P*), or the original home must find a new way to function as a household (*E*, *MP*). The map of their previous or future trajectories, as well as those of their old and new spatial connections is projected and their pace of living is set. The potentially and assuredly linear events accelerate the social and internal time of the characters by filling it up with external agitation as well as with intense internal reflection. The new beginnings, thus, are connected with the new spatial and temporal configurations and inserted in seasonal and social cyclicity.

Winter

The time of intense social activity culminates in the Christmas period, which is joyfully described by pompous Mr. Elton as "quite the season indeed for friendly meetings" (*E* 124). He adds that "at Christmas everybody invites their friends about them, and people think little of even the worst weather" (*E* 124). As in *E*, in *P&P*, *MP* and *P* Christmas is presented as the season of holiday; however, it also marks a pivotal point in the novels, as it is a perceived pivotal point in nature because of the correspondence to the Winter solstice. The Christmas period also corresponds to the first critical point on a psychological level and brings changes to the spatial constellation of the story.

[6] Deresiewicz sees Austen's *Persuasion* as a succession of several two-stage processes such as growth and decay, loss and recovery, desertion and repopulation, parting and reunion, down and up, which, in his opinion, reflect "the deepest movements of nature" (140).

In the beginning of January, and thus, the new calendar year, the spaces in the countryside that had been more or less socially animated during autumnal balls and Christmas gatherings are deserted. In all six novels we see people going away to other places, mostly London or Bath, for the new year means the beginning of the town season. This massive going away is, therefore, part of the social cyclicity of the landed gentry that coincides with the beginning of the new—arbitrary—calendar cycle. In *NA* Catherine leaves for Bath to have her first real-life "adventure," in *S&S* Elinor, even though against her will, and Marianne leave with Mrs. Jennings for London to increase their chances of meeting their loved ones. They are followed by the Middletons with the Steele sisters, as well as their half-brother John Dashwood who must be where the fashionable crowd is; moreover Brandon, Edward, and Willoughby are also in London for the season. In *P&P* the Gardiners go home to London and take Jane with them hoping that the change of place will heal her emotional disappointment. Unlike merely temporary departures, dictated by the social season, Charlotte Lucas leaves for an undetermined period of time (although as her husband is Longbourn's heir they will probably return after Mr. Bennet's death). In *MP* the moving away starts a bit earlier, but in January the Crawfords leave, and Sir Thomas sends Fanny to Portsmouth to teach her a lesson of humility, after she has rejected the offer from Henry Crawford. As in *P&P*, in *E* the Christmas guests leave and the mortified Mr. Elton temporarily escapes the place of humiliation. Finally, in *P* there is the most painful going away—that of Anne, who is obliged to leave her home space and get used to life in Bath. However, several other people from the neighbourhood—family and friends, including Wentworth, will follow her path for various reasons. Thus, the Somersetshire countryside will be left behind for the second half of the novel.

As the main action in *P* unfolds from January on in Bath, the town becomes the linear space, the "point B" and the trajectories of the characters go towards it. Some of them, like Mrs. Smith or even the Elliots, seem to be stuck there for good. In contrast, the Bath of *NA* is rather a cyclical space. It is used as a springboard for various characters that pass through under the pretext of accompanying a sick family member for the opportunities it affords of new socially useful contacts and in order to get somewhere else soon. London is also a cyclical space for the heroines of *S&S* and *P&P* because they go there in pursuit of a personal goal (hoping to meet their love objects and to renew their relationships with them) and not for the place itself. For example, in the case of Marianne, this will go beyond a simple seasonal visit, as she only goes to London because Willoughby is there. Had he stayed in the country, London wouldn't have been of any interest for her. Moreover, this visit will be marked by a traumatizing, thus linear, event. On the other hand, for characters like John and Fanny Dashwood participating in the London season represents a goal in itself. Their country life serves to improve their fortune and give more consequence to their London visits.

The perception of time and space changes for the reader concerning those who stay in the country. We have little indication of what is happening in Wiltshire, Devonshire, or Somersetshire after Catherine, Elinor and Marianne, and Anne respectively leave. Letters connect these spaces with Bath and London, but the reader does not actively witness the events in the country. However, in *P&P* and *E* there is the opposite situation. As the reader is mainly following the point of view of Elizabeth and Emma, and both heroines stay at home during the London/Bath season, we get the other view.

The main plots of the novels, except *P*, are still far from resolved at the end of Winter. For Catherine the great adventure is still ahead, for she has just been invited to Northanger Abbey. Elinor sees Edward's marriage as a certain event, and Marianne has no more hope of married Willoughby. Jane Bennet is still pining for Bingley while Elizabeth's heart is free. Fanny's feelings for Edmund remain secret, and his marriage to Mary Crawford is still possible. Emma is realizing that she is not in love with Frank, while Frank's and Jane's future together is very uncertain. In these novels then Austen presents this Winter period as a time necessary for personal reflection and waiting, with the active social season in the background, before any new elements bring evolution. Actions that do not respect this fallow period do not bring satisfaction, as is the case with Marianne's sending notes to Willoughby in London.

Spring

Spring, in the novels, can be divided between before and after the Easter period. Easter day is usually not mentioned precisely and it does not seem to be important as the Easter period is assimilated into the natural/agricultural year that lies at its origins. March and often the first part of April are characterized by outward movement or outside space: the heroines go away or stay away against their wishes. Catherine is invited to Northanger Abbey and leaves Bath in a state of great anticipation. Elizabeth Bennet leaves Longbourn in order to visit her former best friend Charlotte, freshly married to Mr. Collins, in Kent. Fanny Price is obliged to stay in Portsmouth longer than planned, because the Bertrams are too preoccupied by Tom's health to bring her back. Elinor and Marianne are trying to find means to get home during the month of March, but the first real opportunity to do so occurs only in the beginning of April. Emma stays at home, of course; however, Frank is away from the end of February to the beginning of May.

There is another pivotal point in the plots around the period of Easter or right after it, which mirrors the turning point at the Christmas period. Two traumatizing events in two of the novels happen at this time: Catherine is rudely sent away from the Abbey, and Marianne falls almost fatally ill. The period is marked in *NA*, *S&S*, *P&P*, and *MP* by marriage proposals from the 'right' men or by the right decision not to make this proposal. Henry Tilney and Mr. Darcy propose in mid-April, Edward in the beginning of May. This parallels the natural cycle with the blossoming of Spring. In *MP* Edmund decides not to marry Mary at about this time, a last dying of the unfruitful seeds sown the previous Autumn. Similarly, in *E*, Emma realizes that she is not in love with Frank, and, after his return to Highbury in early May, that he is not in love with her.

The second half of Spring is also the time of returning home in all the novels except *P*. In general, this is the time the London season ends and families begin to return to the country. This movement is evident in the novels, where it sometimes has a more personal and linear aspect. It does not always, that is, correspond to the seasonal cyclicity of social rituals. This is the case of Catherine Morland's return, which is not freely chosen: she is banished from Northanger Abbey and this event will speed up the outcome of the novel. Marianne returns home after the most exhausting months in her life. She is physically recovered, but she will need time to achieve her emotional balance, and she will never really be the same again. Elizabeth and Jane Bennet return after a lengthy stay with their friends and family members. This might look outwardly cyclical as the social year of the gentry was made up of such stays at friends' houses and periods at home. However, it is rather linear inasmuch as Elizabeth returns home completely shaken up by recent events. Fanny experiences an anxious journey home. She still does not know if Edmund will decide to marry Mary Crawford, though good news on that point awaits her at home. Finally, Frank Churchill returns to his original home with secret hopes, not yet known to Emma and the reader. While this return home approximately corresponds to the social cycle of the gentry's return to their country houses after the London season, it also mirrors the agricultural season of going back to working the land (ploughing) after the Winter's break.

Summer and the Other Years

In *P&P* and *E* significant lines of the plot are settled during the Summer and contains the major revelations as I will show in the corresponding sections. In the other novels this season is much less present though it still partly functions as the period of denouement. Finally, the marriages of the heroines take place the following Autumn,[7] most of the time, thus completing the cycle and, at the same time, starting a new one.

Although the main action of the narratives mostly takes place during (approximately) one year, the reader also gets some ideas concerning other years in the lives of the characters which, in a way, frame the main story. However, the size, as well as the importance of this frame is different in each novel, as I will show in the corresponding sections. The "pre-history" of the novels is mostly related by the telling of the linear events that interrupt the presumed social and/or seasonal cyclicity. The parts that succeed the main plots are also not of the same length nor importance in the novels. However, unlike the "pre-histories," the last chapters convey the impression of a newly established social cyclicity for many years to come.

Seasonal Tables

The rhythm of the changing seasons is different in each novel as Tables 2.1, 2.2, 2.3, 2.4, 2.5, and 2.6 illustrate, which show the proportional length of the seasons in relation to the places where the main action of the year happens.[8] I use orange for Autumn, blue for Winter, green for Spring and yellow for Summer. The years and/or months that precede or succeed the main action are coloured in pink.

[7] In his article "Marriage and Birth seasons" published in *Economica* in 1936 Sir William H. Beveridge gives statistics for marriage and birth rates in the 12 months of a year. Although he notes certain differences from one county to another, he concludes that "the seasonal fluctuations" of births and marriages in the mid-nineteenth century are due to social habits (160). Moreover, he adds that "in purely agricultural districts autumn retains much of its importance as a marriage season" (160). This phenomenon is explained by a quotation from William Far (1885) which attributes its causes to "the accumulations of autumn supply a store of food and the harvest wages of young swains" (133). Arguably, Austen's plot patterns correspond to these observed social realities.

[8] The percentages are rounded.

Table 2.1 *Northanger Abbey*

%	3%	45%	48%	3%
Number of Chapters	1	14	15	1
Main action		**Winter**	**Spring**	
Principal places	Fullerton	Bath	Northanger Abbey	Fullerton, Woodston
Other years	Catherine's childhood			Marriage in the following Winter

Table 2.2 *Sense and Sensibility*

%	8%	40%	24%	28%	
Ch.	4	20	12	14	0.5
Main action		**Autumn**	**Winter**	**Spring**	
Principal places	Norland	Barton	London	London, Cleveland, Barton	Delaford, Barton
Other years	Spring, summer				Summer, autumn

Table 2.3 *Pride and Prejudice*

%	53%	5%	25%	16%	1%
Ch.	23	3	15	10	0.5
Main action	**Autumn**	**Winter**	**Spring**	**Summer**	
Principal places	Hertfordshire	Hertford shire.	Kent, Hertfordshire	Hertford shire, Derbyshire	Hert. Derb
Other years					Not indic ated

Table 2.4 *Mansfield Park*

%	7%	17%	25%	33%	16%	2%
Ch.	3.5	7.5	12	16	8	1
Main action		**Summer**	**Autumn**	**Winter**	**Spring**	
Principal places	Mans field	Mansfield	Mansfield	Mansfield Portsmouth	Portsmouth Mansfield	
Other years	8 years					Sum mer →

Table 2.5 *Emma*

%	18%	38%	17%	25%	2%
Ch.	10	21	9	14	1
Main action	**Autumn**	**Winter**	**Spring**	**Summer**	
Principal places	Highbury	Highbury	Highbury	Highbury Donwell	High bury, Don well
Other years					autu mn

The tables highlight an evolution in the seasonal structure of the novels, which I will discuss fully at the end of the chapter. Starting with an equally divided binomial construction in the first novel—*NA*—the structure becomes richer in *S&S* with Autumn and Summer added if in embryo. We find all four seasons present, though unequally in *P&P*, while *MP* and *E* approach more closely seasonal equilibrium. The last novel—*P*—comes back to a simpler seasonal construction. However, the influence of the past, which accompanies the main action of the novel, adds depth to its temporal architecture.

Table 2.6 *Persuasion*

%	4%	13%	38%	41%	4%
Ch.	1	3	9	10	1
Main action		**Summer**	**Autumn**	**Winter**	
Principal places	Kellynch	Kellynch	Uppercross, Lyme	Bath	
Other years		Summer 8 years earlier	Autumn 8 years earlier		Not defined

As we can see from the tables, the only season present in all the novels is Winter. However, as I will show later, it does not carry the same weight in each novel. Its place in the plot also varies: it marks the starting point in *NA*, represents the period of waiting and gestation in the following four novels, and corresponds to the last and concluding part of *P.* Similarly, Summer is the first season in *MP* and *P* and stands as a "prologue" to the forthcoming linear events. In contrast, it is the last season in *P&P* and *E*, and crucial to the outcomes of the plots. This could be explained by the necessity of relating the traumatizing past events that psychologically shaped Fanny and Anne, before exposing them to a new linear trajectory. In contrast, Elizabeth and Emma emerge from rather careless years and are eager for new challenging beginnings in Autumn. Consequently, I have decided to start my seasonal analysis of each novel (except *NA* where this season is absent) with Autumn which, in general, marks the first important linear change in the narratives. I follow, then, the main plot trying to respect the chronology of weeks, months, and seasons as much as possible. At the end of the section I give an overview of the preceding and/or following years in each novel.

Northanger Abbey

If Autumn is often the starting point of the novels' main action, *NA* constitutes an exception. Its main action starts right after Christmas, thus the autumnal months are completely absent. As we can see from the seasonal tables, *NA* has the simplest temporal structure of all the novels. On the other hand, even though only two seasons—Winter and Spring—are present in the main narrative, the fact that Henry Tilney and Catherine get married "within a twelvemonth from the first day of their meeting" (*NA* 261) gives the reader a sense of a completed year.

The timeline of *NA's* main plot starts around Christmas, although this pivotal point is "behind the scenes" in the novel. It must be around Christmas that Catherine, who has been dreaming of romantic adventures for several years, gets the offer from the Allens to go to Bath as they leave in January. The reader also finds out later in the novel that James Morland had spent the second part of his Christmas holidays with the Thorpes (*NA* 25) and his thoughts of marrying Isabella must have been born at that time, which would correspond to pivotal points in other novels. Thus, the significance of Christmas season is already in embryo in *NA*.

The novel's main action commences in Winter months which are spent in Bath, during its high season. In *NA* the cyclicity of the social routine is quickly established: in the morning, going to the shops and the Pump room, leaving one's name in the book; in the afternoon, going to the Upper and the Lower rooms, some nights to dance, and other nights to go to a theatre or a concert. Catherine, however, does not spend much time repeating the routine: after the introduction of "three or four days [...] spent in learning what was mostly worn" (*NA* 12) Catherine meets her hero, Henry Tilney, during the second ball that she attends. The events in *NA* move faster than in other novels. During the six weeks that Catherine spends in Bath, she meets the Tilneys, falls in love with Henry, befriends Eleanor, and gets an invitation to Northanger Abbey. In parallel, she meets Isabella Thorpe, a "false" friend and her brother John, and tries to escape his attentions. As during Frank's visit in *E* or during Elinor's and Marianne's stay in London in *S&S* the time is intensively filled up and almost every day contributes to the development of the novel's action. However, there is a break of three weeks between Catherine's and Henry's first meeting and their seeing each other again, during which she anxiously wonders if she will ever see him again when, that is, she is not absorbed by a gothic novel. She looks for him in the Pump-room (*NA* 23), in the theatre (*NA* 28) and the attentions of the newly met John Thorpe do not make her forget Henry Tilney. We see here in embryo the period of waiting and reflection in Winter that we see more fully developed in other novels.

In the first half of Spring, there are formations of different social constellations in new spaces in all the novels. In *NA* this space is the Abbey—the embodiment of Catherine's gothic dream space (as I discuss in Chap. 2). However, it is also a real family home for the Tilneys; thus, Austen superposes these two spaces: the imaginary and the real. The Abbey is presented as an isolated and closed place. The first part of Spring still being out of the social season in the country, Catherine and Eleanor stay there all the time without any visitors, their only direct connection with the outside world being through letters. As females, they are not free to visit other places on their own which is in contrast with Henry, who goes back and forth between his original and his "professional" home and Frederick, who, as the eldest son, stays or goes wherever he wants. The women also follow a strict timetable imposed by the General. From this point of view, it is a cyclical space: not only because of its ritualized functioning—days are measured by the same activities performed at the same hours—but also because it represents a transitional space for Henry and Eleanor at this time of their lives: Henry already has another home in his parish and Eleanor is to get married and leave sooner or later (being rich, she is certain to find a husband). In Catherine's case, the phenomenon is more complex. She goes to the Abbey as to her "dream house," which would imply linearity. Her experiences during her stay, notably the development of her relationship with Henry, and, especially, at the end when she is forced to leave, are also of linear quality. However, in a long-term perspective, Northanger Abbey, like Bath, is a transitional space/time for her: that is where her relationship with Henry evolves towards marriage, but when married they will live elsewhere.

Like his sons, General Tilney also travels at this time. He leaves home for London where he meets John Thorpe who is not in Bath anymore and whose words will cause the change in the General's behaviour to Catherine. Thanks to his absence, the General gives Catherine a freer time to enjoy her stay, though he calls the time in the country in March or April a "dead time of year" (*NA* 215). He means it in the social sense, bearing in mind that the main activities are going on in London or Bath as the letters from Isabella prove. He also considers that Spring weather is too uncertain for walks outdoors. However, their first outing makes a big impression on Catherine. The building and the space where it is situated are both described: "shut off by knolls of old trees, or luxuriant plantations, and the steep woody hills rising behind to give it shelter, were beautiful even in the leafless month of March" (*NA* 182). Like the description of Barton cottage in *S&S* or Pemberley in *P&P* this gives a sensation of roundness and protective circularity which contrasts with Catherine's wild gothic fantasies about the place and its master, as well as with the General's tyrannical character. This world of unrealistic suppositions and obscure characters are better reflected by the "narrow winding path through a thick grove of old Scotch firs" (*NA* 183) where the girls turn next.

The second half of Spring corresponds to a return home after an emotional shake up in four of Austen's novels: *NA*, *S&S*, *P&P*, and *MP*. However, in *NA* the return is unexpected. Catherine suddenly gets expelled from Northanger Abbey for an unknown reason and her return home is presented as a linear event: she has to leave almost immediately, without any explanation and without any acknowledged possibility of return. Ironically, her solitary and emotionally stressful journey corresponds to the adventure she was previously wishing for. Contrary to what she always expected from adventurous experience, however, Catherine comes back home agitated and unhappy. She is not capable of joining in the household routine and remains in her internal restlessness until the visit of her suitor three days later which "resurrects" her hopes. For Henry Tilney the forced departure of Catherine also represents a linear experience. The reader is used to seeing him as a gentle character who avoids direct confrontation and who accommodates himself to his father's whims, protecting his psychological integrity by adopting an ironic view on the world. However, at this critical moment he openly disobeys his dominating father and proposes to Catherine. He doesn't risk much in material terms, for his part of the inheritance is secured by his parent's marriage settlements (*NA* 259). Obtaining General Tilney's consent remains the only moral obstacle to overcome before they can marry. The main action of the novel ends at this point, leaving this last hurdle in front of them.

The denouement, which starts with the marriage of Eleanor to her titled suitor, takes place in Summer after the main action of the novel. Austen gives details about this surprising event (the reader did not have any information about Eleanor's love life before) at the very end of her last chapter. This advantageous marriage has a positive influence on General Tilney, who accepts Henry as his son again and, thus, removes the last external obstacle for Henry's and Catherine's marriage. This way Summer is in *NA* the time of final resolution.

The main plot of *NA* is preceded by a brief glimpse of the heroine's childhood. The first chapter of the novel, after having resumed the family history in a few sentences, sketches a portrait of Catherine's appearance, character, and accomplishments at the age of ten. From there the narrator points out the transformations that these parameters undergo at the age of fifteen, which can be considered as a linear change, for a little girl is growing up and becoming a woman. Notably, the key ages of both heroines who have their childhood related in the "pre-history" of the main plot—Fanny Price and Catherine—almost correspond: nine/ten, fifteen and eighteen. The introductory chapter of *NA* finishes with the event that will condition the rest of the story: Mr. Allen invites Catherine to accompany him and his wife to Bath. The reader will see a similar, though, more thoroughly elaborated narrative structure in *MP*.

Sense and Sensibility

In *S&S* the first turning point of the main plot takes place at the very beginning of Autumn. The novel, though, starts sometime in early Spring with the death of Mr. Dashwood, which marks the end of a first life-stage for the three Dashwood girls. The next several months of Spring and Summer, during which they continue at Norland, as visitors, and not as mistresses, correspond to a transitional time/space (the gap between the years of living at Norland as their home and the future time of undetermined length elsewhere), up to the moment when Mrs. Dashwood decides to move to Devonshire. This period fits into the four opening chapters of the novel and has an effect of acceleration for the reader who is thus led to consider this as a preliminary to the main story. During this time, too, Norland changes its status as a type of space. From being their home and refuge, it becomes a temporary and open space with a multitude of possible trajectories out of it. Outwardly the Dashwood women are staying there as before: they live through the changes in nature, they are engaged in their usual everyday activities—walks, drawing, and playing the piano—although they have already started their mourning for the lost home, which is inevitable after they have lost the man who was their father and husband. This temporary cyclicity is soon to become the starting point of a very linear movement: they must leave and their departure must be definitive. Once the direction is decided, the cyclicity is broken and the heroines enter the intensive phase of their home-mourning.

Their arrival at Barton cottage marks the beginning of a new life-span far removed from Norland in terms of space and time together with the change of their social status. However, their first impression of the place in this season is favourable: "It was very early in September; the season was fine, and from first seeing the place under the advantage of good weather, they received an impression in its favour which was of material service in recommending it to their lasting approbation" (*S&S* 33–34). The situation of the cottage "surrounded by hills" with a valley in front of it gives the impression of circularity. It is also meant to be a temporary home space: the girls hope to leave it to get married, though their mother might need to remain there for many more years to come. September can thus be seen as a threshold month and it is the end of Summer when the first signs of natural decay appear. This offers an ironic perspective on Marianne's tree mourning scene before quitting Norland. These are the last moments before the Autumn transformation, nevertheless, feeling the imminence of irreversible change in her life, she consoles herself with dreams of eternal unchanged continuity in that particular time and space.

The Dashwoods hardly have time to get settled in their new home when linearity erupts with the force of an accident. Unlike Jane Bennet in *P&P*, who is purposely exposed to the autumnal rain by her mother, Marianne goes out driven by her own fancy. It is the end of September, the weather looks threatening, but Marianne ignores the signs and gets caught by violent rain on the top of a hill. From this very circular and open space Marianne chooses the shortest and the straightest way home: to connect the top of the hill to her garden gate by "running with all possible speed down the steep side of the hill" (*S&S* 50)—a linear and, at the same time, symbolic act, which will, unfortunately for her, bring the most traumatizing consequences. Marianne finishes her journey in Willoughby's arms, thus transforming, in a short moment, her accident into her first love experience. From Willoughby's point of view it is a linear event as well. He usually comes to this place, every year at the same time, to pay a ritual visit to his old relative. He regularly looks at the Barton cottage wishing it were inhabited. Marianne falls in his way while he is returning from hunting, a regular activity for this season. The story with her will change his emotional history, though not as dramatically as for her. The accident takes place just before the autumnal social period of parties and other gatherings organized by Sir John Middleton, which gives the opportunity for Marianne and Willoughby to develop their mutual attachment.

However, the most significant disruption of the cyclicity in *S&S* is the departure of Willoughby; and the drama of this linear event lies in the certainty shared by Marianne and her mother and sister that he will return as Marianne's fiancé. The reader, who follows the story largely through Elinor's eyes, hopes for this return as well, although, aware of the particularities of the genre, the reader will expect some difficulties and obstacles before the lovers can be reunited. Their first meeting having taken place in mid-September, Marianne enjoys the growing intimacy with Willoughby only for a month. From this time on she will go through long months of suffering, which will only end with her recovery after Easter.

However, despite the absence of Willoughby and Colonel Brandon, the period of autumnal social gatherings continues in its cyclical regularity. It brings disappointment for Elinor too, when she learns about Edward's prior engagement, though, in her case, it will turn out to be a temporary suspense of her hopes and she will receive his marriage proposal in May, after the period of wintry anticipation, once she is back at Barton cottage.

As in *NA*, the Christmas period in *S&S* is not a part of the narrative. However, as in the other novels, it does represent, if indirectly, a turning point. This must be the moment ("on the approach of January" [*S&S* 174]) when Mrs. Jennings invites Elinor and Marianne to go with her to London and participate in its season with its opportunities to meet Willoughby and Edward Ferrars. This helps to advance the action which seems to be blocked at that point, Willoughby and Edward being gone for unknown reasons and for an undetermined time. Christmas is also mentioned as a "festival which requires a more than ordinary share of private balls and large dinners to proclaim its importance" (*S&S* 173), thus, the culminating point of the social season.

As in *NA*, *P&P*, *MP*, and *P*, January takes the heroines away from home. Even though in general, going to London for "the season" is a ritual for a large part of the gentry, for Elinor and Marianne, it is an unexpected event. London is distant from their cottage, the journey takes three days, so they are far removed from home in spatial and temporal terms. Mrs. Dashwood stays with Margaret in the empty cottage and the almost deserted environment, while the older girls move into the very centre of fashionable social life. Nevertheless, the first three weeks after their arrival are fairly eventless, especially for Marianne, who is waiting in vain for Willoughby. They follow the usual social cyclicity by visiting people and getting visited by them, but that gives little satisfaction as the ones they really want to see are absent or out of their reach. Despite the ongoing social activities, this is for Elinor, as for Emma, Elizabeth, and Fanny, a time of observation, reflection, a certain "digestion" of the situation. More precisely, Elinor is mourning Edward and the hopes she had concerning the future of their relationship. At the same time, she is observing Marianne's emotional involvement and trying to guess, without daring to ask, the seriousness of her sister's relationship with Willoughby.

For Marianne, on the other hand, this is the time of suspension before another "accident": the ball, where she accidently sees Willoughby with another woman and he refuses to acknowledge their relationship. This happens at the end of January and it is the pivotal point for Marianne: the event is deeply traumatizing and she will never again be the person she used to be. Here we see again linearity disrupting the cyclicity of London's seasonal balls. This time, Marianne's accident is public: mirroring her first fall in Autumn she "falls" again, fortunately on a chair, while her sister "expecting every moment to see her faint, tried to screen her from the observation of the others" (*S&S* 202). Thus, at this instant not only is the linear confronted to the cyclical, but the private is confronted to the public. The cyclicity of seasonal activities goes together with the public space: huge public gatherings held at this time of the year in a city space are shared by many members of the genteel society. The linearity, on the other hand, corresponds to the private: the personal history that evolves in this public and seasonal space/time. In a way, both Dashwood girls cannot fully participate in the London season: their minds are too occupied by their personal problems. At first, Marianne goes out expecting to meet Willoughby, and no other person or event interests her. After the "break up" she closes herself in her internal space/time and loses almost all contact with the outside world. Elinor controls the outward expression of her emotions and acts in a publicly acceptable manner, but inside she is mourning her own and her sister's relationship.

At the end of Winter, however, Elinor is confronted to another emotional "accident" which has direct consequences for her life. Edward and the Steele sisters arrive in London in February, and at the beginning of March the "accidental" slip from Anne Steel reveals Edward's and Lucy's secret engagement. This speeds up the resolution of the situation, which was doomed to last unchanged for years till Mrs. Ferrars' death. And as Edward refuses to break the engagement, there seems to be no hope left for a first-time reader as well as for Elinor. Thus, at this point the narrative gives a false impression of an end and makes space for a surprising event in the future.

The beginning of Spring for Elinor and Marianne is still the time of suffering. After the two weeks' stay with her daughter who gave birth in late February, Mrs. Jennings is back home with both Dashwood girls to continue with their usual routine. However, Elinor is plunged into the problems caused by Edward's and Lucy's engagement and, moreover, is forced to act as an intermediary between Edward and Colonel Brandon, who is offering him a living. Both girls feel they have been far from home for a long time and wish to be away from London, this strange and publicly exposed space, where they have seen their expectations crushed:

> The Miss Dashwoods had now been rather more than two months in town, and Marianne's impatience to be gone increased every day. She sighed for the air, the liberty, the quiet of the country; and fancied that if any place could give her ease, Barton must do it. Elinor was hardly less anxious than herself for their removal … (*S&S* 316)

Paradoxically, two months earlier Marianne wished to be away from Barton with the same intensity, and the only place that could "give her ease" was the one where she could be with Willoughby. After the disappointment she associates London with her sorrow and attributes to the town space her emotional confinement. Her wish for an open-countryside space expresses the need to liberate her mind from that oppressive emotional pain.

However, circumstances do not allow them to go straight home and they must pass through Cleveland, the Palmers' home, where they go for the Easter holidays. Geographically, it brings them closer to Barton at the distance of "a long day's journey" (*S&S* 317). For Marianne, however, this place is associated with Willoughby, because his house is in the same county though thirty miles away. Situated on high ground, Cleveland offers an open view to the countryside and Marianne fancies that she can almost see Willoughby's house, Combe Magna (*S&S* 343). Being a transitional place on their journey home, Cleveland acquires a linear quality in terms of space and time insofar as it hosts the most dramatic scene of the novel. Marianne falls seriously ill in this house, close to where her life with Willoughby should have taken place, and, symbolically, during the period of Easter. Easter Sunday is not clearly referenced in the novel, but the reader is informed earlier that the Palmers go to Cleveland for the Easter holidays. Physically, Marianne mirrors the miracle of the resurrection: after seven days of crises she, surprisingly, gets over her illness. Mentally it will take a much longer time for her to mend: a year and a half will pass before her marriage to Colonel Brandon. While she is slowly recovering her physical strength, important adjustments take place out of her earshot. Willoughby visits Cleveland and tells his side of the story to Elinor, while Colonel Brandon confesses his feelings for Marianne to her mother and thus gains Mrs. Dashwood's appreciation. From this culminating point in space/time the narrative will take a well determined direction with respect to Marianne's story line: Willoughby will disappear from her life for good and she will gradually get closer to Colonel Brandon.

Thus, the heroines of *S&S* return home in late April after a series of trying linear events and expect to find their inner peace there. During their two days' journey Elinor notices in Marianne the signs of her internal transformation:

> she, who had seen her week after week so constantly suffering, oppressed by anguish of heart which she had neither courage to speak of, nor fortitude to conceal, now saw with a joy, which no other could equally share, an apparent composure of mind, which, in being the result as she trusted of serious reflection, must eventually lead her to contentment and cheerfulness. (*S&S* 387).

Elinor sees the qualitative change in her sister's personality: in the future Marianne's happiness will not spring from spontaneous uncontrollable emotions but will come from calm reflexivity. In other words, in order to reach serenity, Marianne has to abandon her way of perceiving life as a chain of linear, thus destabilizing, events and experiences, which almost killed her. However, the very day after their return to the cottage, the reader notices that this peace of mind is not that easy for her to achieve. By deciding to follow an excessively strict self-imposed routine, Marianne risks falling into another extreme: "I mean never to be later in rising than six, and from that time till dinner I shall divide every moment between music and reading" (*S&S* 388). She also promises to live only for her family and never go away from home (*S&S* 393). Her project is not in harmony with the natural cyclicity of life: it is a self-imposed, rigid cycle and so unlikely to succeed.

At this point the narrative leaves Marianne on her winding pathway towards accepting human life's cyclicity and turns towards the unresolved fate of Elinor. Fortunately for her, soon after having heard the false news of Edward's marriage, she is visited by him and gets engaged. Similarly to Catherine and Henry in *NA* Edward proposes very soon after Elinor's return home, before she even has a chance to get settled—as if to say that she does not need to settle there again, and when she no longer expects such a resolution of the situation. This element of surprise seems to show that the characters have to accept the contingency of events as well as social and natural cyclicity. In fact, this Winter-Spring trip to London and back represents Elinor's emotional quest: she passes through the most trying stages, has to give up all her hopes and definitely renounce the man she loves in order to gain him on her return home. As in the previous novel, the only obstacle that remains is external: the consent of the parent—Mrs. Ferrars in this case—and some additional money to live on. At the end of May Edward visits his and Elinor's future home, the parsonage in Delaford and ritualistically "accepts" their new home space. He also manages to obtain the consent of his mother at the same period. Thus, by the end of Spring the main action of the novel is almost completed and a map of the new family constellation is drawn. In consequence, Summer in the novel is summarized in its last chapter as the time of waiting for their new home—the Parsonage of Delaford—to be ready.

The last chapter of *S&S*, which is constructed in a similar way as the last chapter of *P*, hosts one more linear event—Marianne's change of heart and marriage to Colonel Brandon (*S&S* 429), which, however, was prefigured as a possible outcome after Marianne's illness.

Pride and Prejudice

P&P starts with a fresh event late in September: "Netherfield Park is let" at Michaelmas and this causes the excitement and expectations of new connections, preferably culminating in a marriage or two. As we can see from the seasonal tables, Autumn is the most prominent month in *P&P* occupying a little more than half of the novel. The pace of the narrative is slow, with numerous dialogues and descriptive pauses, creating an effect of deceleration for the reader whose attention is thus focused on this period of time. The season is full of new arrivals that will initiate linear events: Mr. Bingley, Mr. Darcy, the militia regiment with Mr. Wickham and Mr. Collins all arrive in the county. Noticeably, all four newcomers will marry Hertfordshire ladies within a year. The time of the narrative is filled with walks, balls, parties, and visits that generate new, strong, and, in Elizabeth's and Mr. Darcy's case, contradictory emotions in the characters, as well as in the reader. Austen sets the principal plot lines in Autumn, which she develops and resolves during the following seasons. Thus, Autumn in *P&P*, symbolically emphasizes the new, potentially fertile seeds being sown and it is full of youthful energy and optimistic projects for the future.

Like all but one of Austen's heroines (Catherine Morland), the Miss Bennets get their share of disruptive linearity later in Autumn. However, in contrast to *S&S*, where the departure of Willoughby in late October is definitive, in *P&P* the departure of "the lover" is temporary, though the Bennets, as well as first-time readers, are not yet aware of it. This disruption is provoked by the most significant event in *P&P* at this season—the Netherfield ball late in November. The time that precedes the event is entirely filled with expectations of pleasure and advancement in marriage prospects in Longbourn as the girls and their mother must stay at home almost the whole week because of "a succession of rain" (*P&P* 98). This episode illustrates once more the dependence of females on the weather. Nevertheless, this period of confinement is psychologically necessary for the heroines: this is the time of projecting, anticipating, and dreaming of the future. The reader, on the other hand, experiences an acceleration of pace at this point and is thus led to expect some major events at the ball. However, the ball turns out to be a disappointment for the reader as well as the characters, and Netherfield Park finishes its short season by not being the centre of attraction for the neighbourhood for another year to come: the Bingleys and Darcy leave and the Bennets have to look for social interaction and marital expectations elsewhere. However, two of them remain focused on this abandoned space: Jane and her mother. For Jane it represents a certain memory-space, where she was happy in the company of the "most amiable man" (*P&P* 152), while Mrs. Bennet shifts backwards and forwards from despair to hope of making it Jane's home. The regiment of militia with its officers brings a longer lasting revival of Longbourn's everyday life and longer lasting hopes for change. However, the socializing with the officers is depicted as a very superficial activity never producing any serious marriage projects. It mainly creates a new place of attraction in Meryton, one more place for entertainment in the social map of the neighbourhood. The stay of the regiment is temporary, although the regularity of the journeys to and from Meryton and Longbourn makes it a part of the everyday cyclicity.

As in other novels, Christmas in *P&P* is also a moment for a large family gathering: the Gardiners usually visit Longbourn at this time of the year. It is the first moment in the novel that the reader meets the family members that Elizabeth is proud of who will also have an important role in her relationship with Mr. Darcy. Here the period of Christmas also represents the point of maximal intensity in the family's social life: in one week "there was not a day without its engagement" and "Mrs Bennet had so carefully provided for the entertainment of her brother and sister, that they did not once sit down to a family dinner" (*P&P* 161). This sentence has an ironic touch, for even though Christmas is a socially active period, the first goal of the Gardiners is to visit their family members in Longbourn; thus, family dinner would be most suitable. However, it is Mrs. Bennet's wish to prove her social standing that determines the agenda.

Mrs. Gardiner is an alternative maternal figure to whom the events of the Autumn are told and who gives the most "objective" opinion. She assesses Elizabeth's relationship with Mr. Wickham, warning her about the potential "impossibility" of the match. In this way, Christmas marks the turning point in Elizabeth's and Mr. Wickham's growing intimacy: it is made clear that their relationship cannot evolve any further. Moreover, the marital hopes connected with the Netherfield family and with the officers lose their probability at this time; the excitement of the previous autumn has given place to disappointment and Mrs. Bennet bitterly relates to her sister-in-law that "two of her girls had been on the point of marriage, and after all there was nothing to it" (*P&P* 159). In general, Christmas at Longbourn marks a transitional period. It puts an end to several expectations (as mentioned earlier), removes certain characters or gives them a different status and, thus, allows the narrative to develop. This is the case with Charlotte's marriage: by marrying Mr. Collins she erases the possibility of the Bennets' continuing at Longbourn for another generation and, at the same time, by leaving with him she opens a connection to a new space that can and will bring new opportunities.

During the rest of Winter Elizabeth stays at Longbourn without the two people closest to her: Jane and freshly married Charlotte. And though both women write letters relating their disappointments and their joys, "with no greater events than these in the Longbourn family, and otherwise diversified by little beyond the walks to Meryton, sometimes dirty and sometimes cold, did January and February pass away" (*P&P* 171). As we can see from the seasonal tables, Winter is the shortest season in the narrative—it takes only three chapters or 5% of the novel. Moreover, one and a half of the Winter chapters are dedicated to the Christmas period that I have analysed. The reader, thus, is given very little time to follow this gestational period: time is accelerated till the following major event—Elizabeth's trip to Kent in March. On the other hand, the condensation of January and February into one sentence serves to highlight their two most distinctive features—cold and dirt. This gives the impression of an eventless, lonely time/space of hibernation framed by unfavourable weather conditions. Moreover, Elizabeth, at this time, is reflecting more on others' emotional state than her own. Jane is sad because she has not seen Mr. Bingley and because she is finally convinced of Caroline's desire to drop the acquaintance. Charlotte, on the other hand, seems joyful, though Elizabeth doubts her sincerity. At this time Elizabeth also realizes how little she was in love with Mr. Wickham, for he starts courting another woman and she does not feel any pain. Thus, during the months of Winter she is also gradually freeing her heart from this supposed attachment, and, unconsciously, preparing for more important psychological experiences that she will go through in Spring.

As in *NA*, Spring in *P&P* begins with Elizabeth Bennet being invited (though indirectly) to a great house in March—Rosings "well situated on rising ground" (*P&P* 178). She comes to Kent to visit her friend Charlotte, who lives in the neighbourhood of Rosings after her marriage to Mr. Collins, and stays there for five weeks. As indicated earlier, such lengthy visits to friends were common amongst the gentry: people, considered as genteel, would alternate their stay at home with long visits to friends and family according to the season and their financial resources. Elizabeth has no possibility of participating in the London season any more than her sister Jane who is in London, but staying with her tradesman uncle who does not move in the "right" circles. For Elizabeth the visit to Charlotte is a pleasant alternative to the social season in town. Thus, outwardly it resembles a usual, cyclical activity, whereas inwardly she is about to live a startling experience which will have linear consequences.

During the visit Elizabeth's time is also divided in a ritualistic cyclical way: there are morning visits, daily walks, as well as dinners twice a week and evening parties with Lady Catherine. The arrival of Mr. Darcy and Colonel Fitzwilliam brings some variety to the routine, but it still stays more or less regular up to the crucial moment of Mr. Darcy's proposal. Like Catherine and Henry during her visit to Northanger Abbey, Elizabeth and Mr. Darcy have this time to get to know each other better. In Mr. Darcy's case these days have "fatal" consequences: he falls in love and will propose right after Easter. His regular Spring visit to his aunt will turn out this time to be his last. Elizabeth, on the other hand, thanks to the information from Colonel Fitzwilliam, becomes more prejudiced against him than ever. We might say that she is at the opposite emotional pole to love. Symbolically, their emotional responses to each other, though not yet of the same quality, grow from the seeds sown in the previous Autumn during their first encounters.

Retrospectively the reader may notice that the dinner at Rosings on Easter Sunday marks the pivotal point in Mr. Darcy's attitude towards Elizabeth. From this evening on he visits Elizabeth at the parsonage "almost every day" (*P&P* 202) and these visits culminate in the marriage proposal. The most important confrontation (and the last of an aggressive kind) between Elizabeth and Mr. Darcy takes place at that critical moment.

This "explosion" will make them look back and re-evaluate their memories, the bases of their opinions, their beliefs, and their deeper selves. They experience a real linear internal transformation and live their introspective moments of self-examination out of measurable time. Moreover, at this point, and regardless of Elizabeth's refusal, the "impossible" relationship starts to appear possible to the reader.

In *P&P* the reader has what is a unique opportunity in the novels to observe the evolution of two characters, namely Charlotte Lucas and Mr. Collins, before and after their marriage. In other novels, we only hear about a character's past as single because they get married before the start of the main action of the novel, or else they get married at the end of it. Mr. Elton in *E*, might be considered an exception, but the reader does not meet his wife before they get married and he brings her to Highbury. Charlotte Lucas becomes Mrs. Collins right after the Christmas festivities (on January 9) and moves into a different stage of her life. The reader can observe, through the eyes of Elizabeth, the way she organizes her daily routines, the places in the house that she has appropriated, and the ritual visits to her husband's patroness and the parish. Charlotte's joy and pleasure in having Elizabeth with her, and her strategies for forgetting her husband as often as possible show that her new life is not based on marital felicity. We see her enjoying these newly established cyclical activities, but we read that "her home and her house-keeping, her parish and her poultry, and all their dependent concerns, had not *yet* lost their charms" (*P&P* 240 emphasis mine). With this gently ironic statement Austen seems to suggest that this type of household-centred cyclicity does not bring lasting happiness.

In the second half of Spring *P&P*, like *E*, is still a few months away from the final resolution of its plot. Elizabeth and Jane Bennet come back home in mid-May. Being at home and having again some privacy, allows Elizabeth to tell Jane her news about Mr. Darcy: his proposal and his letter. Their return is also mirrored by the return to Longbourn's initial status of a calm and dull place, as it was just before the start of the main action of the novel. As the regiment is about to leave for Brighton and Mr. Bingley has no plans of coming back to Netherfield Park, their home space feels quite empty. The girls fear the return of the usual household cyclicity without any novelty from the outside world. This makes Lydia fight to follow the officers and, once she has obtained the invitation to Brighton and its "campfull of soldiers," (*P&P* 243), Kitty cries out of jealousy. It also incites Elizabeth to look forward to her Summer trip with the Gardiners with much enthusiasm and it is "her best consolation for all the uncomfortable hours, which the discontentedness of her mother and Kitty made inevitable" (*P&P* 263). On the other hand, time spent at home, away from intense social activity, allows Elizabeth (and the reader) to understand the deep feelings that Jane still has for Mr. Bingley (*P&P* 252) as well as to partially relate her own concerning Mr. Darcy and Mr. Wickham to Jane. Thus, as in *S&S*, *MP*, *E*, and *P*, the forced confinement is fertile internally as it allows psychological advancement through self-examination.

Arguably, after the new beginnings in the first Autumn, the gestation period full of doubts during Winter, and the new information in Spring that makes germination possible, Elizabeth, Jane, and Lydia approach the denouement of their emotional situations in Summer. At the very end of May Lydia, full of enthusiasm, leaves for Brighton with the regiment and hopes "to enjoy herself as much as possible" (*P&P* 261). Her trip, as well as Elizabeth's later on, will turn out to be a linear event, not only for herself, but also for all her family. In general, this departure mirrors the departure of the Bingleys, and then, Jane with the Gardiners in Winter giving an impression that Longbourn and those who remain there are left behind. This sensation of emptiness is confirmed by Elizabeth's thoughts in which she projects herself towards the future lake-tour in order to fill up the socially empty present after the departure of the regiment (*P&P* 263), as I discussed earlier. However, as this is the countryside season for the gentry, Longbourn re-establishes its social connections when "the families who had been in town for the winter came back again" (*P&P* 264). Elizabeth, however, will have this seasonal and social cyclicity disrupted by another trip, organized by her uncle during the time of his Summer holidays. At this point a different social cycle, that of a tradesman working in town and taking a few weeks annual break when his business allows him, collides with the traditional social cycle of the English landed gentry and their pleasurable season in the country that lasts for several months. Elizabeth accepts to be part of this different social cycle for three weeks and this will bring her to take a decisive linear step towards the most radical change in her emotional life.

The tour, which is shortened twice, goes through Derbyshire allowing Elizabeth to get gradually acquainted with her future husband's home-county, and ends, though a few days earlier than planned, in Lambton, near Pemberley, his home. The last place Elizabeth visits is Pemberley, and the last person she talks to before the precipitate departure is Mr. Darcy. The tour and its final destination reflect Elizabeth's emotional journey towards knowing and loving Mr. Darcy. The Summer season is also significant, as it corresponds to her maturing feelings and understanding of her own real inclination. Mr. Darcy, at the moment of their meeting in Pemberley, has already gone through important internal changes, has evaluated his previous behaviour, and overcome his social pride. Thus, their mutual disposition, which gradually evolved during Spring, begins to yield visible and hopeful "fruit" in the beginning of August.

The destiny of Jane will evolve as well. However, unlike her sisters and similarly to Fanny Price (in *MP)*, she will stay passive and faithful to her feelings for her only love, Mr. Bingley. Thus, while Elizabeth is going through the qualitative interior change and physically moving from one place to another, Jane, always steady, is passively waiting at home for the change of external circumstances. Moreover, her steadiness is vital to the Longbourn family during the most trying period—Lydia's elopement with Mr. Wickham.

As in *E* the last part of *P&P* brings its three single females to their marriages the following Autumn. Lydia appears to be the most "gifted" for breaking cyclicity at the first opportunity that she gets. After having participated in the social cycle for hardly a season (the accepted age of "coming out" was fifteen at the time[9] and Lydia is sixteen when she gets married [*P&P* 337]) and after having flirted, though never seriously, with several different men, she—surprisingly—elopes with Mr. Wickham and indirectly forces her family and friends to go looking for her and to organize her marriage. The marriage takes place hurriedly, but not entirely out of the usual season for such occasions—at the end of August. As in other cases of forced linearity discussed earlier, Lydia's impulsive act doesn't bring her a happy future. In the last chapter of the novel Austen gives a few details of their unsuccessful married life qualifying their life style as

[9] Genteel girls used to be introduced into society at 17 or 18 (Davidoff 51), however, in some families the younger sisters had to wait till the older ones were married before they could come out (*MP 651* note 7) Lady Catherine has this practice in mind when she is questioning Elizabeth about her family in Vol. II Ch. 6 (*P&P* 187) and referring to the Miss Bennets being out together as "very odd." However, we notice that most of the families Austen writes about do not stick to this rule (the exception is the Lucas family in *P&P* where after Charlotte's engagement "the younger girls formed hopes of coming out a year or two sooner than they might otherwise have done" (*P&P* 137): in *S&S* the two Dashwood sisters are out together (Margaret is only 13, thus too young); in *P* it must also have been the case for the youngest sister, Mary, was the first to get married; in *MP* the Miss Bertrams are out together and only Fanny has to wait till she is 18 (however, when Fanny's coming out ball was organized, Maria Bertram was already married); the Thorpe sisters in *NA* are also out together.

"unsettled in the extreme" and their affection as short-lived (*P&P* 429). The reader can thus perceive an implied authorial disapproval concerning the non-respect of natural and social cycles. On the other hand, Lydia's elopement gives a significant push to the growing intimacy between Mr. Darcy and Elizabeth. Only when thinking that she has lost him for good, Elizabeth realizes how important he has become to her; likewise, seeing Elizabeth vulnerable and her respectability threatened leads Mr. Darcy to act regardless of his principles and save the possibility of marrying her.

Summer in *P&P* ends with Lydia's marriage. However, the final denouement of the story takes place in Autumn. In *S&S*, on the other hand, there is the unwritten story of Marianne and Colonel Brandon that takes another year to develop, and in *MP* we never find out how long it takes for Edmund to propose to Fanny. In contrast, in *P&P* the period from the return of Mr. Bingley and Mr. Darcy for the autumnal shooting in Netherfield to Mr. Darcy's second proposal is recorded almost day by day. Mr. Bingley proposes during a weekend in late September, possibly straight after Michaelmas, which would correspond to almost exactly a year after their first meeting. About two weeks later Mr. Darcy proposes to Elizabeth, which would again correspond to their first meeting in mid-October (during the assembly ball in Meryton) the previous year. Their marriages must take place later in the same Autumn, for Elizabeth invites the Gardiners to Pemberley for Christmas (*P&P* 424). Moreover, as in *E*, there is a child soon to be born that more or less mirrors the emotional evolution of the main heroine. Mr. Collins announces "Charlotte's situation and his expectation of a young olive-branch" (*P&P* 403) in his last letter to Mr. Bennet. Bearing in mind that their marriage took place in the beginning of January, the child would be due in Autumn, thus starting a new cycle of human life and bringing Charlotte to a different phase of her life—that of being a mother.

The last chapters of the novel convey the impression of a newly established social cyclicity for many years to come. For example, in the last chapter, Mrs. Bennet is said to visit Mrs. Bingley and talk of Mrs. Darcy with "delighted pride" (P&P 427). Thus, the reader gets a picture of this social cyclicity: her regular visits to her neighbours and her repetitive conversation turning around the same subject—her advantageously married daughters, especially as she is shown to have a habit of going over and over the same topics at several instances during the novel. Likewise, the cyclicity of the new life for other characters—Lydia and Mr. Wickham, always short of money and regularly taking advantage of Jane's situation;

Mr. Bennet and the Gardiners with their repetitive visits to Pemberley; Georgiana, Kitty, and Mary Bennet gradually growing into women—are outlined. However, the reader learns nothing about Elizabeth's and Mr. Darcy's relationship to each other after the marriage—their couple is presented as a part of their social circle and their future lives are sketched in relationship to the other members of the same circle. This situation, however, has generated quite a few sequels such as Shannon Winslow's *The Darcys of Pemberley* and P. D. James' *Death Comes to Pemberley.*

Mansfield Park

In *MP* the autumnal linear change comes slightly later than in the other five novels. October rain prevents the hunting (in this case men also are prevented from outside activity by the weather) and serves as an excuse for the inside "theatricals," which will rearrange the constellation of the protagonists by giving a more important place to Fanny in her uncle's eyes and by causing a complex chain of events. This imaginary space/time stands out of the cyclical everyday life; it also promotes attachments between the participants some of which will have dramatic (!) consequences later. The return of Mr. Bertram at this time accelerates his daughter Maria's calculated marriage. She gets married to Mr. Rushworth in November, after having fallen in love with Henry Crawford earlier that Autumn. In contrast to Austen's heroines, Maria tries to overcome her newly born passion for Henry and follow the pre-established plan of an advantageous though loveless union. Her behaviour can be seen as unnatural, for, symbolically, her matrimonial seeds sown this Autumn are not fertile and will not be able to thrive. On the other hand, Maria's marriage means her removal from her childhood home. Her sister Julia goes away leaving Fanny the only young girl in the house, which will expose her to Henry's attentions. At the same time Edmund starts developing serious feelings for Mary Crawford, with a pending "deadline" of his ordination in the Christmas period, all the more tense as Mary believes herself unable to live with a clergyman.

The Christmas period in *MP* is socially and emotionally important for Fanny. She makes her first official dinner visit, even if only to the parsonage; her brother William, the closest person she has, visits her; and Henry Crawford starts flirting with her, even though she does not immediately realize it. But the culmination of her introduction to the social world is the ball given by Sir Thomas on December 22 especially for her. The culminating event is that Fanny dances (*MP* 293)—significantly the first and the last time that we see her dance in the novel—and thus "enters" adulthood. It is also the only ball that is organized at Mansfield Park, consequently, the moment is clearly linear. For Edmund it is a pivotal point too, "his mind being deeply occupied in the consideration of two important events now at hand, which were to fix his fate in life—ordination and matrimony" (*MP* 296). The day after the ball he leaves for his ordination, which will take place during Christmas week, whereas his intended bride Miss Crawford is still resolved on not marrying a clergyman. The actual Christmas week passes "quietly and peaceably at the great house" (*MP* 331) where, in contrast to Uppercross in *P* or Longbourn in *P&P*, Fanny is the only young person.

However, for Mary Crawford, Christmas is the moment of an important decision, which will determine her future. She could cut the cyclicity of her long stays at other peoples' houses—her sister's, the Admiral's, and her friends'—always hoping to meet a convenient match, by marrying Edmund and becoming a country clergyman's wife. Being "confined within doors by a series of rain and snow" (*MP* 332), she almost physically feels the distance from Edmund and misses him as he delays his return to Mansfield. Her altered emotional state makes London, her chosen affective home-space, seem unbearable (*MP* 333). Like Emma examining her mind in snowed up Hartfield (discussed later), Mary in the Parsonage, goes through her memories, regretting "what she said" and wondering about the future:

> If she had felt impatience and regret before—if she had been sorry for what she said, and feared its too strong effect on him, she now felt and feared it all tenfold more. […] But at any rate his staying away at a time, when, according to all preceding plans, she was to remove to London, meant something that she could not bear. (*MP* 333)

Mary feels hurt by Edmund's choice of distance at this particular time and blames her own behaviour. Consequently, the only space that she could inhabit without fear and sorrow would be the one where Edmund is. This is a common device in Austen's fiction to place these self-examining soliloquies in the intimacy of real or substitute home spaces; as well as Mary Crawford and Emma there are similar scenes with Anne in the empty Uppercross house (*P* 133), and with Marianne Dashwood who pours her heart out in her last letter to Willoughby in a "gloomy morning in January" in London (*S&S* 205). Looking closer, we notice that the characters do not freely choose to be in this place, but are confined to their rooms, because of the bad weather. The gloom, rain, or snow seem to fall at the right moment, as if supporting their need for solitary pondering. Thus, we could say, that this natural, "meteorological" cyclicity (the ongoing succession of rain, snow, and sunshine according to the seasons) supports the linear transformation of the characters on the psychological level by allowing their reflections.

The holiday period in *MP* ends with an unexpected turn for Fanny: Henry Crawford's marriage proposal, which she rejects. However, as neither Henry nor Sir Thomas give up their efforts at making her change her mind, the situation is left suspended. In general, as in *E* and *P&P*, it is an important time that brings together several linear events. January is, moreover, centred on this important linear event: Fanny's rejection of Henry Crawford. Sir Thomas tries to convince her to change her mind, Lady Bertram and Mrs. Norris show their disapproval and only Edmund, though negatively surprised, takes time to understand Fanny's motives. While her family is, for the first time, concentrated on her, Fanny herself is closely watching the evolution of the relationship between Edmund and Mary, which, unfortunately for her, appears quite favourable at that time. Thus, January in *MP* is also a month of reflection, contemplation, and suspension in accordance with the natural cycle.

There is, however, a significant qualitative change in February. The reader follows Fanny in her exile to Portsmouth, where she loses control over time and space. She spends three months there, which bring psychological turmoil. Like Bath for Anne Elliot, it is a place of banishment for her, though, unlike Anne, she knows she won't need to stay there for years, but only an unknown number of weeks. Fanny's visit is not the equivalent of a seasonal visit to Bath or London. Portsmouth represents a linear space *par excellence*. She is sent there by her uncle to undergo linear changes in her mind. In the opinion of Sir Thomas, this stay in her ancient home space should break her determination, change her attitude, and transform her feelings. Comically, events will rather transform his own attitude towards Fanny and modify his evaluation of her decisions.

The Portsmouth visit is a trying and transforming time for Fanny. She begins to be dependent on people's letters for keeping track of what is going on at home and in London where Edmund, the Crawfords, Maria, and Julia are. Moreover, she tries to guess how soon her friends might write and where they might be at a particular moment. Even the letters from her rival Mary become very welcome to Fanny and her thirst for contact with her usual world that springs from a feeling of vulnerability and the fear of abandonment surprises her:

> Here was another strange revolution of mind!—She was really glad to receive the letter when it did come. In her present exile from good society, and distance from every thing that had been wont to interest her, a letter from one belonging to the set where her heart lived, written with affection, and some degree of elegance, was thoroughly acceptable. (*MP* 455)

This extract indicates that though physically Fanny is in Portsmouth, psychologically she is still invested in her adoptive home space. We could even consider the space "where her heart lived" as a virtual home: it does not only mean Mansfield Park, for Edmund is in London at this time and she follows him there emotionally. A letter is experienced as a piece of emotionally comforting space/time extracted from the world where she is still internally dwelling. In addition, Portsmouth being a seaside town has a liminal quality which symbolically parallels Fanny's liminal psychological state. She experiences Portsmouth as a disorienting space on the border of the sea, which represents the unknown. Her brother William, with whom she has developed a close emotional relationship, only rapidly passes through, to get onto a ship and sail away, leaving Fanny alone and lonely in her former home. Paradoxically, staying there helps her cut all psychological ties with the space and time of her early years. Therefore, it is a linear space for Fanny: the point she was sent to and which she wishes to leave as soon as possible. It is a linear experience in terms of time too, although it can be qualified as passive linearity. She won't need to change her mind by deep reflection: events in the outside world will show that she was always right and in this respect, she is in a very different position from Emma or Elizabeth. Actually, her status in Mansfield Park will change for the better after her return.

Similarly to Elinor and Marianne in *S&S*, Fanny is obliged to stay longer than planned in a place—in her case her former home—where she feels trapped and lonely. She relies on letters from Edmund, Mary, and her aunt to keep the connection with the world to which she feels she belongs. Even Henry Crawford's visit before Easter is not as unwelcome as his visits to Mansfield Park used to be and his departure makes her feel more isolated than ever: "it was parting with somebody of the nature of the friend; and though in one light glad to have him gone, it seemed as if she was now deserted by everybody; it was a sort of renewed separation from Mansfield" (*MP* 479). Fanny dreams of going back to her Mansfield life. From a larger perspective, she never wishes for any real change, but strives to remain in her cyclical ways. Her love for her cousin Edmund, with whom she grew up, reflects this desire to remain in a familiar space. Being away makes her feel as if a part of her life was cut off. Fanny imagines Spring in Mansfield, which she is forced to miss this year:

> She had not known before, how much the beginnings and progress of vegetation had delighted her.—What animation both of body and mind, she had derived from watching the advance of that season which cannot, in spite of its capriciousness, be unlovely, and seeing its increasing beauties, from the earliest flowers, in the warmest divisions of her aunt's garden, to the opening of leaves of her uncle's plantations, and glory of his woods.—To be losing such pleasures was no trifle … (*MP* 500)

This passage in which the narrator's voice blends with the character's shows how much Fanny is connected to her home space, and how much her psychological well-being is conditioned by being close to the ever-changing nature. It is important to note that Fanny realizes her attachment to Mansfield Park only when she is away from it. Tellingly, she represents to herself the images of nature together with the memory of her aunt and uncle: it is her aunt's garden that is so beautiful and her uncle's woods that are so glorious. They are contrasted with the "confinement, bad air, bad smells" (*MP* 500) of her former home space. Thus, by evoking the missed rebirth of nature, she draws a picture of her emotional landscape. Fanny's thoughts contrast with Marianne's tree mourning scene at the beginning of *S&S* (discussed earlier). Fanny is focusing on the natural change of the season, which she is contrasting with Spring in Portsmouth where she is cut off from nature, and the inevitable transformation in nature that takes place without her being able to observe it, while Marianne imagines everlasting Summer in her childhood home. This reflects the capacity of Fanny to grow psychologically with the natural flow, to feel comfortable in the cyclicity of life, and Marianne's early determination to fight and resist it, before she is transformed by her illness. On the other hand, it also shows their different relationships with their childhood homes and their dissimilar family histories.

While away, Fanny becomes a passive spectator of linear events in her adoptive family which take place without her participating in them or having any direct influence on them: Tom's illness, Maria's and Julia's elopements and, finally, Edmund's decision not to marry Mary Crawford. Arguably, Fanny's trip to Portsmouth, her absence from the Bertrams is necessary as, after these destabilizing events, they can seek for her soothing, steady presence. By being away and passive she, paradoxically, accomplishes a linear transformation: that of gaining an important place in the Bertram family.

The end of the Easter holidays is the limit that Sir Thomas sets for Fanny's return to Mansfield which vexes her: "Not till after Easter!—How shall I bear it?" (*MP* 491). As in other novels, Easter events here bring a definitive turn in the narrative. During the Easter holidays Henry meets Maria in London and renews the flirtation. They run away soon after, followed by Julia and Mr. Yates. This immediately liberates Fanny from her guilt at refusing Henry in the eyes of the Bertrams, and emotionally detaches her from the weighty friendship with his sister. This will also lead Edmund to renounce his project of marrying Mary Crawford and, thus, free him for Fanny.

Finally, at the beginning of May, Fanny Price returns home with silent and distressed Edmund, who has come to fetch her. However, it takes time for Fanny to get back into the usual everyday cyclicity of Mansfield Park. Maria's elopement, which meant the loss of a daughter to Sir Thomas and Lady Bertram, requires time to be processed and Fanny does her best by being an attentive listener. However, she has to wait for several days after their arrival at Mansfield till Edmund is finally ready to talk to her about his meeting with Mary Crawford and his decision not to marry her. At this point, at the end of Spring and after an emotionally exhausting chain of events, Edmund is freed from obsessive feelings for Mary, as Marianne Dashwood is freed from obsessive feelings for Willoughby. He will start getting affectively closer to a "calmer" option—Fanny, who will, in time, be "rewarded" for her patience. He will need time for his brotherly feelings to turn into sexual attraction, as Marianne needs time for her perception of Colonel Brandon to change from that of a father-figure to that of a husband. As in *S&S* the end of Spring marks the end of the novel's main action, even though no marriage is announced yet.

Summer in *MP* has some importance on the psychological level. This is the time for Edmund to recover from his unsuccessful love story with Mary Crawford and transfer his feelings to Fanny:

> After wandering about and sitting under trees with Fanny all the summer evenings, he had so well talked his mind into submission, as to be very tolerably cheerful again. […] Scarcely had he done regretting Mary Crawford,

> and observing to Fanny how impossible it was that he should ever meet with such another woman, before it began to strike him whether a very different kind of woman might not do just as well—or a great deal better; whether Fanny herself were not *growing* as dear, as important to him in all her smiles, and all her ways, as Mary Crawford had ever been. (*MP* 535, 543–544, emphasis mine)

The narrator's intention to "purposely abstain from dates on this occasion" (*MP* 544) means that we cannot know when exactly this change takes place, nor when they actually get married. However, as the previous citation shows, Summer is the period of hopeful waiting for Fanny (as opposed to the period of helpless and rather painful waiting in January), as it is for Elinor Dashwood and Catherine Morland, and a time of emotional transition for Edmund. Regular and repetitive conversations with Fanny during Summer look like a cyclical experience. Nevertheless, it gradually transforms his emotional state and finally turns out to be linear on a psychological and practical level. Fanny does her job just by being there, as she did in Spring by *not* being there, and listening. The word "growing" is also significant for it reflects the natural processes of Spring and Summer: it is his feelings for Fanny that are growing and maturing at this time.

Moreover, the Summer months that precede the first autumnal changes in *MP* are also important for the narrative. The Crawfords arrive in July (*MP* 46) and bring the emotional stir into the Mansfield household. Thus, the forthcoming linear events germinate and start growing during the last part of Summer, which does not exactly correspond to the seasonal cyclicity. Bearing in mind that all the romances connected with Mr. and Miss Crawford end in disillusionment, it is symbolic that they start at an "unnatural" moment in time. On the other hand, the Summer heat is an important aspect of this period: it facilitates the emotionally reckless acts and stimulates sexual desire, as for Maria during the visit to Sotherton (in a passage I analyse closely in Chap. 3). Edmund's feelings are also affected by the season, although in a less "violent" way. During this first Summer, which takes a significant 17% part of the narrative, he starts falling in love with Mary Crawford:

> A young woman, pretty, lively, with a harp as elegant as herself; both placed near a window, cut down to the ground, and opening on a little lawn, surrounded by shrubs in the rich foliage of summer, was enough to catch any man's heart. The season, the scene, the air, were all favourable to tenderness and sentiment. (*MP* 76)

The scene is described as an artificial, deliberately arranged composition, displayed in a window and beautifully framed by the luxuriant green. It looks as if its positive effect has been calculated by Mary, whose "unnatural" character the reader discovers progressively. The sentiment in Edmund is also "false" for it rises from the accumulation of external elements that strengthen each other's effects. Significantly, as if to illustrate a contrast between false and true love, his authentic feelings for Fanny also appear during Summer, a year later, although in this case they rise from close friendship and mutual confidence.

In *MP* Austen accords more attention and place to the pre-history than in her other novels. Arguably, this is crucial to understanding the circumstances and conditions of Fanny's arrival and her first years in the Bertram family and it is related in material and psychological detail in the first four chapters. The novel starts with the advantageous marriage of Maria Ward to Sir Thomas Bertram thirty years before the main action, which lays the foundation for the narrative and determines the future relationships between the Ward sisters. This is followed by the adoption of nine-year-old Fanny Price several years later and her painful integration into the household of Mansfield Park. The next linear event related is the death of Mr. Norris and the possibility of change. Fanny, who is fifteen at the time, fears that she will have to live with her aunt Mrs. Norris, which shows how much she already feels at home at Mansfield by that time. And, finally, the reader gets acquainted with the events that set the direct background to the main action: the trip of Sir Thomas to Antigua and the arrival of the parsonage guests—the Crawfords.

On the other hand, the closing chapter in *MP* conveys a sense of steady cyclicity in the relationships between Fanny, Edmund, Sir Thomas, Lady Bertram, Susan, and William. Moreover, there is an impact on an almost closed family circle: people that Fanny cares for are gathered around her after their acquisition of Mansfield parsonage. The exception is her favourite brother who is not physically close to her, but he is advancing in his career as Fanny wanted him to do. In general, this mirrors *P&P*'s epilogue in which the account of future years is concentrated around Pemberley.

Emma

As in *S&S*, *P&P*, and *P*, the main action in *E* starts in September with Miss Taylor's marriage leaving Emma in her sad wondering "how was she to bear the change?" (*E* 5). This phrase anticipates the constant search for balance between the usual—the cyclical—and change—the linear indicating a new beginning for the heroine. Emma has no wish to leave Hartfield and her father, even for a short visit to London, but she cannot be satisfied without engaging her mind in a new project. Nevertheless, standing on this threshold of September Emma anticipates the struggle through the long evenings of October and November with only her father for company (5). In the small Highbury society, autumnal months bring lonely seclusion at home without the advantage of some private or public balls to break the routine. For this reason, Emma is obliged to look for a new personal project to survive psychologically and takes a young girl of an unknown family—Harriet—under her wing.

The Autumn months following Miss Taylor's marriage are then filled with a certain hierarchy of projects. Emma starts by finding her new companion—Harriet Smith—and then begins "improving" her *protégée's* life by drawing her attention to Mr. Elton, who is the principal "object" of this period. Hartfield is the central place and all the action turns around it at this time. Outwardly Emma sticks to the regular and the usual: long walks with Harriet, drawing, visiting neighbours and the poor, collecting charades, and talking with Harriet about her love affairs. Secretly, however, she nurtures an ambitious linear project—to get Harriet "well" married. Comically, the real linear events are developing in parallel to the imagined (and real!) love signs from Mr. Elton without Emma's awareness: Harriet's sincere affection for Robert Martin and her own more than friendly feelings for Mr. Knightley as well as Mr. Elton's "false" feelings for Emma. Thus, the fact that Emma induces Harriet to reject the marriage proposal from Mr. Martin will not have any linear consequences: it will take a year, but the following Summer Harriet and Robert will be as much in love as they first were. Interestingly, at this very period another real love story is developing outside the main narrative: Frank Churchill meets Jane Fairfax in Weymouth and they get secretly engaged in October. Their relationship evolves hidden behind the main plot, and except a couple of (for a first-time reader hardly perceivable) hints and Mr. Knightley's suspicions in June, it gets dramatically disclosed only in July.

In *E* there is the most elaborate Christmas event in all the novels: the unfortunate Christmas Eve party which takes place on the mythical night of the Winter solstice, the lowest point before the (re)birth of the natural astronomical and Pagan-Christian cycle. At the same time, the event exposes and reveals the characters of the participants to the reader. The characters that make up this party know each other well enough and some of them at least should be capable of predicting each other's reactions to any unexpected situation. The exception is Mr. Elton who is the only relative "stranger" in the company. Nevertheless, there is some novelty in this traditional Christmas gathering. It is the first relatively big party organized at Randalls after Mr. Weston's marriage and it is the first time that the new Mrs. Weston is a hostess in her own home instead of filling the role of an ambiguous companion at Hartfield. Mr. Woodhouse, Isabella, and John Knightley are going to this newly formed household to "inaugurate" the new place and discover how warm Mr. Weston's fires are.

Half a mile separates Hartfield from Randalls, but, ironically, the journey reveals itself as full of danger on the physical and psychological levels. The cold and snow, which correspond to the common image of the longest night of the year, comically become a serious problem and only Mr. Elton seems to be delighted with the idea of being "snowed up a week at Randalls" (*E* 124). For Mr. Woodhouse it is one of the two times that he stirs from Hartfield during the novel's main action, the other time being in Summer, the day before Midsummer, which corresponds to the summer solstice—exactly six months later—when he visits Donwell, during the strawberry picking party that I will analyse later. For the first he surprises everybody by not changing his mind even when it threatens to snow (*E* 121). There seem to be some key moments in the natural cycle that Mr. Woodhouse is determined to honour whatever the circumstances! On the other hand, in one of his exuberantly pessimistic moods John Knightley sees the trip as the "defiance of the voice of nature, which tells man, in every thing given to his view or his feelings, to stay at home himself, and keep all under shelter that he can" (*E* 122). His description of the short trip to the neighbours' house appears more appropriate to a dangerous expedition through uninhabited northern lands. Still, despite the natural obstacles the guests arrive at their destination, and the only two people "excused" from the compulsory Christmas expedition are (for Emma and the first time reader) the supposed bride-to-be sick Harriet and the ghostly Frank Churchill. Thus, luckily for Mr. Elton and his real matrimonial intentions, both possible human obstacles are out of the way. The journey to Randalls, with the exaggerated danger of a thin layer of snow, is presented like a real linear event. Ironically, it turns out actually to be one for Emma.

The journey home, after the party, is even more psychologically trying than the arrival, especially for Emma. At the end of the evening all the guests are in emotional turmoil because of the supposed danger of not being able to get home safely. The only two able to keep their heads are

Emma, who has spent the evening trying to escape Mr. Elton, and George Knightley. For the experienced reader, this is a sign of the shared level headedness of Emma and Knightley, which hints at their real, though not fully acknowledged complicity. At the same time this contrasts with Emma's "false" relationship with Mr. Elton, which is approaching its critical point at the end of the party. On the journey back, Emma is confined to the closed space of the carriage with Mr. Elton and the ride seems to her unbearably extended in time. After his marriage proposal she is forced to face her own blindness: the past, the present, and the future unite in this restricted and non-anchored carriage space. While crossing the distance between Randalls and the Parsonage, and, later on, between the Parsonage and Hartfield during the longest night of the year, she is crossing the first border between her naive fantasies and reality. However, before the end of the novel she will have to cross several other borders of this type. For Harriet and Mr. Elton, on the other hand, this moment marks a radical change: Mr. Elton knows he must look elsewhere for a bride and Harriet has to abandon her hopes of marrying Mr. Elton, even though she doesn't know it yet. The narrative will take a different direction from this turning point and new plot lines will be developed. The reader, aware of Emma's misjudgement of the situation, will take a step back from her point of view, and become more attentive to "objective" details or try to read "through" her perceptions.

That night, alone in her room, Emma goes through her memories, tries to comprehend them in respect to the present, and dreads the future, when she will have to tell everything to Harriet. She realizes her misjudgements; however, she is no dupe as to Mr. Elton's real feelings and motives. This extended moment is framed in a recurring everyday situation (as the use of the definite article highlights)—"the hair was curled, and the maid sent away" (*E* 145)—but Emma is facing the disintegration of the project she has been working on during the autumn. In the privacy and intimacy of her bedroom space she can take time, during the longest night of the year, to reorganize her thoughts, suppositions, and resolutions and accept this linear change of the situation. Christmas Day itself brings good news for Emma giving her some additional time to deal with her emotions: snow keeps her at home for several days as well as others—the exception being Mr. Knightley who shows and declares his wish for unrestricted mobility several times through the novel. This holiday period ends then in a closed family circle, without any threatening or disturbing influence from the outside.

As in the other novels, there are several departures after Christmas in *E*: the Knightleys return home to London, Mr. Elton goes to Bath, while the Campbells leave for Ireland. However, all these places remain secondary compared to Highbury, which stays at the heart of the novel—it's the only centre of any foreground action, because Emma never leaves it and the narrative is mainly constructed from her point of view. Highbury is thus opposed to all other places, which are therefore "outside."

January in *E* is a still more socially empty period than Autumn. After Christmas, and before the arrival of Frank Churchill in the first half of February, there are hardly any events. The time seems to be filled with Emma's relating to Harriet the true intentions of Mr. Elton. As a result, Harriet is mourning her disappointed love while Emma is digesting her guilt in giving her false expectations (Vol. I, Ch. 17). The next chapter brings the second event that needs to be discussed and comprehended: Frank postpones his visit once again. This leads to Emma's and Mr. Knightley's discussion about duty and amiability which seems to fill the rest of the month and ends the first volume. This situation echoes the state of nature in the middle of Winter, the period of calmness, reflection, and rumination being equivalent to the season of hibernation. In a way, January in Highbury can be called not only a natural, but also an emotional Winter. For the reader these last chapters of the first volume have a similar wintery effect of slowing down as the textual pause corresponds to an emotional as well as plot pause between the resolution of the Elton-Harriet "project" and the new plot lines that will be opened by the arrival of Frank Churchill. At the end of the month, however, Jane Fairfax arrives—an unusual event, for "normally" she should be enjoying herself at London balls. This starts the new, secretive line in the story, which will cause the whole chain of linear events as the year advances. Unlike *P&P*, Winter is the longest season in *E* taking up 38% of the novel, and containing two busy, eventful months, December and February, with the socially empty and almost eventless slow January in between.

Frank Churchill finally visits Highbury in February moving the frozen situation, satisfying Emma's curiosity and making her forget all the unpleasant past stories with Elton the news of whose engagement arrives at the end of January. What is not emphasized in the novel at this moment is the almost simultaneous return (at an interval of one or two weeks) of Jane Fairfax and Frank, who are both, it is worth adding, children sent away from home and raised by people other than their parents. Jane comes home at an unusual time, under the pretext of her health, whereas Frank visits his native village for the first time since he left it. In general, the arrival of Frank slows up and consequently charges the space/time of the novel for the reader: almost every day is recorded and every meeting of the characters is detailed. Emma goes through two weeks of intense social activity and intense flirtation with Frank. Comically, as Emma cannot go to town for the season, the season embodied by the fashionable and loud Frank, comes to her. This time, contrary to the relationship with Mr. Elton, she believes herself to be the object of admiration and even imagines the marriage proposal from Frank: suddenly she starts behaving as a young, marriageable girl, as at twenty-one she actually is, and not as a match-making, interfering "aunt." She obviously enjoys the break of the everyday cyclicity, which culminates in an improvised dance at the Coles and in the planning of a real ball. All these linear events happen around Valentine's Day (already established as a day for lovers [notes on *Emma*, 572]), and though not stated by the narrator, it has been calculated that Jane gets her mysterious piano present on this precise day, which can thus be described as one of the well-kept secrets of the novel.

However, Emma's February flirtation will turn out to be short-lived. Frank leaves unexpectedly before the end of the month:

> It was a sad change. They had been meeting almost every day since his arrival. Certainly his being at Randalls had given great spirit to the last two weeks [...]. It had been a very happy fortnight, and forlorn must be the sinking from it into the common course of Hartfield days. (*E* 282)

Here Emma is in a mood similar to the mood she was in after Miss Taylor's marriage in September: she is dreading the routine, lack of novelty and of projects. Noticeably, the pleasure is destroyed by the sudden departure of Frank: after the reflexive tranquillity of January that Emma accepted with serenity, February brings novelty and excitement, which, however, is unnaturally and prematurely ended. Jane Fairfax is even more affected: "her composure was odious," "she had been particularly unwell" (*E* 283). First-time readers, like the other inhabitants of Highbury, do not know the real reason for her unhappiness yet. In a way, Jane's feelings parallel Emma's at this time, though hers are stronger and deeper. Jane seems to have entered a linear chain of events and nothing that happens in her life at this time is for her cyclical. She is officially in a transitional period before entering a professional life, and, in that way, partly renouncing her claims to gentility. Privately, she is in a transitional period too: even though others don't know it, she is waiting for a denouement of her relationship with Frank and, thus, hoping never to be obliged to work as a governess. In contrast, Emma is not upset for a long time after Frank's departure, the very next morning she is "busy and cheerful" (*E* 284), and soon she is convinced of not having been in love at all. She is looking forward to Frank's possible return in Spring and starts building a new marital project for Harriet: in other words, she is back to her usual mode of psychological functioning.

As indicated earlier, Winter in Highbury ends with the announcement of Mr. Elton's marriage. This is a hurried, speedy marriage, for he has known his wife for four weeks only before their engagement. Another character who gets married in February is Willoughby, and he too has personal reasons to accelerate his marriage to a fortunate woman. Both marriages are uncommonly precipitated and mercenary, thus "unnatural," which is signalled by their failure to fit into a natural, social, and liturgical cycle.

Frank returns at the end of Spring, or, more precisely, he frequently comes and goes from Richmond to Highbury. The key event of this period is the long-planned ball at The Crown Inn, which finally takes place at the end of May. A ball is an exceptional event in Highbury and therefore can be qualified as a linear experience. It is held not at somebody's home but in in a more neutral space which, moreover, is decorated and transformed as if with "Aladdin's lamp" according to Miss Bates (*E* 348). This transformation of an old inn into a "fairy land" (*E* 349) is matched by some important relational transformations that take place in this space during the ball. Harriet changes her love object from Mr. Elton to her "saviour" Mr. Knightley; Emma is definitely convinced that Frank is not in love with her; and Mr. Knightley and Emma dance together for the first time. There is a qualitative change in their relationship summed up by the words of Mr. Knightley after Emma (!) asks him to dance: "Brother and Sister! no, indeed" (*E* 358). The real linearity of the ball is closely followed by the false linearity of the episode of the gypsies, when Frank helps Harriet to get away from the gypsy children (vol. III, ch. 3) and gives Emma one more chance to be misled by her imagination. Emma considers this moment as the beginning of her friend's love for Frank. The reader, however, is only partly convinced: Harriet's description of the "service" is vague and Emma has already lost her infallibility as an observer. Nevertheless, a new confusing constellation is generated which will develop during the beginning of Summer up to the final conclusion in July.

Austen emphasizes the beginning of Summer in *E* by summarizing the situation of the characters and by leading the reader into some doubts concerning Frank's and Jane's relationship via the thoughts of Mr. Knightley (*E* 372). However, these suspicions do not find any response and/or change of behaviour in Emma for the moment and on the Sucklings postponing their visit Highbury society is back in its cyclical routines. All the principal characters seem to be in want of a new project at this point. The tension between the usual and the new will return with Mr. Knightley's strawberry picking party followed by the excursion to Box Hill. The visit to Donwell takes place on the 23 June, thus the trip to Box Hill is on Midsummer Day (which is 24 June according to pagan tradition). Both days are characterized by the heat, which gives them a slow pace and a certain nervous tension, comparable to electric tension in the air before a thunderstorm. It is also the opposite point to Christmas in the natural as well as liturgical cycle: the winter solstice's heavy snow and cold are counterbalanced by the extreme heat of the Summer solstice. Comically, the reader also gets a glimpse of a Midsummer bonfire: Mr. Woodhouse requires one in the chimney of Donwell. As I mentioned earlier, Mr. Woodhouse seems to be particularly attached to this astronomical/liturgical calendar or rather to the two principal sociocultural milestones of the year. It is the second time in the novel that he stirs from home, the first time being for the Christmas party at Randalls, as we have seen, which gives him a very regular and slow rhythm of "every solstice." Moreover, the two parties that he gives at Hartfield—for Miss Taylor on her wedding day, which must have taken place a little before Frank's letter of congratulations dated 28 September (*E* 109), and for Mrs. Elton as a bride (*E* 314)—approximately correspond to the Autumn and Spring equinoxes. Thus, we see the anxious Mr. Woodhouse functioning according to the reassuring cyclicity of the sociocultural and natural calendar.

The visit to Donwell Abbey follows a plan: arrival, strawberry picking, removing to a shade, walking and admiring the views, going into the house to eat, and walking back out to the gardens. This renders a rhythmical and cyclical experience, which is in relation to the heat. This is in contrast with Jane's and Frank's hectic behaviour towards the end of the afternoon. First, Frank is not there when he should be, then Jane Fairfax insists on escaping the party and walking home on her own. This incident is followed by the actual arrival of Frank, who looks irritated and out of humour. It must be evident for the reader that the two of them do not share the same calm, enjoyable time as the other characters. However, even though it is clear that the cyclicity of the experience is disrupted, a first-time reader cannot know why. Moreover, Frank expresses his wish to travel, to break away from his cyclical everydayness. This is misleading for a first-time reader and it stops the reader making the connection between Frank and Jane. An experienced reader, on the other hand, can see Frank's speech as an emotional response to Jane's act, as well as their loss of faith in a favourable outcome to their engagement. A wish to go away is a wish for linearity, of going from one point to another and never turning back, of plunging into strange spaces that might bring new experiences. Psychologically this represents Frank's desire for the denouement of the situation that seems to be indefinitely stuck, his desire to break the time of anticipation and secrecy that has become exhausting for them both.

By telling Emma that she is "wearied in spirits" (*E* 394) and that the biggest kindness would be to let her go, Jane expresses a similar desire without declaring it. Even though her escape is limited to a few solitary wanderings, she shares Frank's inner state, the difference of expression being due to their different tempers and possibilities. Frank is a wealthy heir and can allow himself to make such travel plans, whereas for Jane such plans would be impossible. The same (economic) reason will later push Jane to accept the offered position of governess. She knows she must go through a linear change at this point of her life as she cannot afford to continue in the cyclical life style of the English gentry, while waiting for an opportunity to change her status to that of a married woman. Thus, the series of disputes and misunderstandings with Frank that lasts from 23 June into early July pushes her to accept the radical change and end the intolerable period of anticipation.

In general, Donwell Abbey is depicted as a realm of cyclicity with its gardens, meadows, flocks, and fish-ponds under a bright, but not oppressive English midsummer sun (*E* 391). We get the sense of comfortable and plentiful auto-sufficiency as well as seasonality, though, as several scholars have noticed, Austen errs in mentioning an "orchard in blossom" at the end of June (*E* 391). However, this seasonal (deliberate?) error reinforces the effect of the landscape on the characters and on the reader. The image of the blossoming trees that can be seen from far away represents a new, fertile cycle. Moreover, with the image of the clover to be cut on the morrow (*E* 392), Austen plunges the reader into the agricultural world where everything is in rotation according to the seasons. The description of Donwell and its grounds also marks a pause in the pace of the narrative that allows the reader to get the sense of the place.

The panoramic view that opens in front of the visitors is vast, however, it is still experienced as a familiar, home-like space:

> a broad short avenue of limes [...] led to nothing; nothing but a view at the end over a low stone wall with high pillars, which seemed intended, in their erection, to give the appearance of an approach to the house, which never had been there. Disputable, however, as might be the taste of such termination, it was in itself a charming walk, and the view which closed it extremely pretty. (*E* 391)

The view that "closes" the space gives it the sense of roundness and the impression as if it still was an inside space. By placing Frank's inclination to go abroad in this context, Austen creates a contrast between the cyclicity of homelike space, which she clearly favours, and the linearity of Frank's desire to escape, which sounds almost heretical at this point in the narrative and in the presence of this marvellous "home" landscape.

On the next day, Midsummer, the trip to Box Hill is organized. Like the Christmas party that was full of disturbances and that ended with a violent experience for Emma and Elton, this day is not pure enjoyment: "there was a languor, a want of spirits, a want of union, which could not be got over" (*E* 399). The focus is on personal relationships overtly or secretly hostile and confusing. The place itself can be perceived as an outside space of linear quality that hosts linear acts and decisions. It is indeed the only place that Emma visits outside Highbury during the whole of the novel, which marks it out as linear. Like the Christmas party, it ends with an emotional jolt for Emma after she is reprimanded by Mr. Knightley and, as the reader finds out later, for Mr. Knightley himself. At this point Emma experiences another inner crisis of self-re-evaluation:

> She was vexed beyond what could have been expressed—almost beyond what she could conceal. Never had she felt so agitated, mortified, grieved, at any circumstance in her life. She was most forcibly struck. The truth of his representation there was no denying. She felt it at her heart. [...] How could she have exposed herself to such ill opinion in any one she valued! And how suffer him to leave her without saying one word of gratitude, of concurrence, of common kindness! (*E* 409)

The expression "forcibly struck" represents the figurative storm towards which the heavy atmosphere of the two days has been tending. This will turn out to be the moment of real transformation, for Emma fears the irreversible consequences of her behaviour. Noticeably, she is mostly struck by Mr. Knightley's opinion of her, which shows her having feelings for him she does not yet understand, rather than by the unkindness of her words to Miss Bates. This conflict will speed up the resolution of the relationship between Emma and Mr. Knightley. For him, it is the critical moment, because, as he acknowledges later, it is at this point that he realizes his love for Emma and his jealousy towards Frank. He chooses to go away. However, this time a solitary walk to Donwell is not enough. He must move further to a different space, not connected with Highbury, and stay away longer. Emma, on the other hand, takes the opposite decision: to get closer, in a physical and psychological way, to people she had hurt in the past. Mr. Knightley is in the linear, he tries to "break up," while Emma tries to mend what she thinks she has broken, and to bring things back to their cyclical mode.

July is the month of resolution in *E*. The death of Mrs. Churchill speeds up the action of the novel by pushing it towards the end and in a few weeks the social map of Highbury is redefined as a consequence of four important changes. This process starts with the information about Frank and Jane's secret engagement. As in the cases of Elinor Dashwood, Catherine Morland, and, partly, Fanny Price, after the decisive exterior change, there will be a few months of serene waiting before the marriage can take place in November: the mourning time for Mrs. Churchill must be respected. Retrospectively, the reader finds out that their relationship started in Weymouth the previous September, which corresponds to the other new beginnings in the novels, but the gestation period was more secretive than in other cases, although they had to go through the same moments of doubts and hopelessness as all other couples during Winter and Spring.

The second major change is connected with Emma herself. Though embarrassed at her behaviour to Frank and Jane, and vexed because of Frank's secrecy and dishonesty, she is much more affected by the news concerning Harriet's hopes for Mr. Knightley which makes her suddenly realize her own feelings for him:

> A few minutes were sufficient for making her acquainted with her own heart. A mind like her's, once opening to suspicion, made rapid progress. She touched—she admitted—she acknowledged the whole truth. […] It darted through her, with the speed of an arrow, that Mr. Knightley must marry no one but herself! (*E* 444)

This moment of recognition is described as something physical: the truth is touchable and piercing. The image of a darting arrow represents the linearity of the moment. There is also a focus on rapidity and immediacy. The cyclicity of her life-long relationship with Mr. Knightley is cut at this moment and there will be no way back to the former mode, even though Emma, when doubting his feelings for her, will wish for such a return to the past. During the following two days Emma goes through and re-evaluates her relationships with Harriet and with Mr. Knightley. She clings to her old idea of never getting married in order to protect herself from the possible disappointment. At this point the tension between the cyclicity—the life before the recognition of her feelings—and linearity—the supposed life after Mr. Knightley has married Harriet—is at its strongest point. Cyclicity seems to be the secure option:

> She had no hope, nothing to deserve the name of hope, that he could have that sort of affection for herself which was now in question; but there was a hope […] that Harriet might have deceived herself, and be overrating his regard for *her*.—Wish it she must, for his sake—be the consequence nothing to herself, but his remaining single all his life. Could she be secure of that, indeed, of his never marrying at all, she believed she should be perfectly satisfied.—Let him but continue the same Mr. Knightley to all the world … (*E* 453, emphasis in the original)

Emma's wish to fix her world, which is on the threshold of definite transformation, springs out of her fear of an unfavourable change. This reminds us of Marianne's tree mourning scene in *S&S*, discussed earlier. The difference lies in Marianne's conviction that Norland's trees will stay unchanged, while she is away, whereas Emma is conscious of wishing an impossible thing. This desired return to cyclicity represents to her an imaginary space/time of refuge. The gloomy Summer days (*E* 459) bring back to her mind the evening after Mrs. Weston's marriage. Emma is dreading solitude once again, however, this time, as before, her fears will turn out to be unfounded. Her engagement to Mr. Knightley together with his proposition to live at Hartfield will move her onto another stage in her life and, at the same time, give her all the hope of being able to continue constructing her cyclical social world without being obliged to move away from her home space, which would normally be the case for married women.

The third major change in Highbury is the birth of a little baby-girl at Randalls. Mrs. Weston's pregnancy is mentioned for the first time in the novel at the beginning of July (*E* 460), right before the actual birth in the second half of July. At this point, the reader sees this forthcoming event through Emma's eyes as a part of the pessimistic picture she draws when fearing future loneliness. However, an experienced reader might notice the change in Mrs. Weston's social behaviour during Spring, which hints at her being pregnant. She refuses to dance when asked during the Crown ball, she stays with Mr. Woodhouse in the house while others are picking strawberries and walking in the heat, and she doesn't take part in the Box Hill excursion organized by her husband. Significantly, other characters do not talk about her pregnancy and she does not discuss it even with Emma. Nevertheless, being pregnant means a change in her socializing habits: she starts turning in her "personal" cycle of maternity connected to the cycle of human life. Bearing in mind the frequency of mother-baby deaths at the time it is a psychologically tense and anxious moment: on the one hand, the mother is participating in the perpetuation of the life cycle, but, on the other hand, she and her baby are very close to possible death. This general anxiety about giving birth is confirmed by Austen in chapter 17, which starts with the announcement of the birth: "Mrs. Weston's friends were all made happy by her safety; and if the satisfaction of her well-doing could be increased to Emma, it was by knowing her to be the mother of a little girl" (*E* 503).

Moreover, this birth symbolically accomplishes the cycle that started in September with Miss Taylor's wedding and, thus, the change in Emma's life. The baby must have been conceived right after the wedding, developed during the gestation period of Winter and Spring to be born in July. This mirrors the natural and the agricultural cycle of seeds sown in Autumn and giving fruit in Summer. At the same time, it parallels Emma's eye-opening trajectory of development. From new unsuccessful projects with Harriet and Mr. Elton and later with Frank Churchill, she moves towards understanding of her real feelings for Mr. Knightley. Their engagement is announced to their family and friends right after the birth of the baby (*E* 508). Comically, Mr. Knightley, though sixteen years older than Emma, was as much in the dark concerning "his own heart" as her. His awakening starts in Winter with the arrival of Frank: "he had been in love with Emma, and jealous of Frank Churchill, from about the same period, one sentiment having probably enlightened him as to the other" (*E* 471). However, the pivotal moment occurs during the Box Hill party, on Midsummer Day, when he chooses to go away and try to "escape" his feelings. Basically, the evolution of their self-understanding is similar in terms of time.

The last important news that comes to Highbury in the middle of August is the engagement between Harriet Smith and Robert Martin. At the beginning of her friendship with Emma, Harriet is in love with Robert Martin, however, under Emma's influence she goes through different transformations in her emotional life: hope and disappointment with Mr. Elton, followed by hope and disappointment with Mr. Knightley. Actually, with her marriage in September Harriet completes her personal cycle. After having first rejected the marriage proposal she accepts it a year later and, thus, gets back to her starting point. The difference is that instead of changing her social status and entering another stage of her life at seventeen she does this at eighteen and after some painful emotional experiences.

The marriages of the three single female characters of the novel take place in Autumn: in September, October, and November. This again situates the new beginnings in Autumn, thus starting a new cycle in natural/agricultural, personal, and social terms, for once again there are new seeds to be sown and a different life, as a married woman, to be constructed.

Persuasion

Like Netherfield Park in *P&P*, Kellynch Hall in *P* is rented at Michaelmas, so Anne Elliot, like the Dashwood girls, is forced to leave home in September without any choice of place or time, and obliged to get accustomed to a different kind of life. She experiences temporary, transitional spaces at Uppercross, at Lady Russell's house, and, later, in Bath. Sir Walter and Elizabeth can at least choose the direction of their exile and they leave for Bath with newly born hopes to re-affirm their status in society. The arrival of the new tenants at Kellynch raises expectations comparable to those raised by events in *P&P* and they also bring some spectacular linear changes.

The second linear event in Autumn is the arrival of Captain Wentworth in search of a new beginning in October and ready to fall in love with some attractive girl. This event starts a new cycle in Anne's mourning of her lost love: she has spent eight years of reflection in Wentworth's absence, and now she is going to have the opportunity to face her emotional state in Wentworth's presence. At the same time, she gets a new chance to experience her love and repair the damage. There is a sense of a return, but on a different level: eight years later they are not the same people. Anne dwells on the change brought by this time apart and Wentworth says he finds her "so altered he should not have known" (*P* 65) her. Their meetings take place in a different spatial configuration too. Wentworth is staying at

Kellynch, thus occupying the space formerly occupied by the girl who rejected him—in other words, he "invades" her territory. Anne is driven away from her childhood home and finds herself in a weakened transitory position. They mainly see each other in crowded spaces in the Musgrove houses, which is, in addition, a temporary space for Anne, who is spending her last months in the country before the definite departure. At this point the novel does not seem to push the action in one particular direction. There is a lingering sense of cyclicity: we do not know which trajectory Anne and Wentworth will take and, at the same time, we get a strong feeling of return, as well as of gradual, but non-definite transformation which evokes seasonal change. On the other hand, Anne is plunging into her feelings with a desperate melancholy, contrasted and spiced with jealousy towards Louisa and Henrietta. It is significant that she cannot concentrate on the beauty of the decaying November landscape and "the last smiles of the year" (*P* 90): her psychological period of decay has lasted for eight years, and she is more than ripe to re-engage in a love story. In a way she is following two different cycles at the same time. In terms of the change in her life circumstances she is starting a different social cycle that mirrors the season, whereas in terms of her emotional situation she is finishing a very long phase of mourning that has been keeping her in an "autumnal" mood for years. Linda Bree sees the signs of Anne's approaching rebirth during the November walk to Winthrop:

> As the action of the novel moves from summer to autumn in the early chapters, it seems as if Anne's fading hope of a fulfilled life will die with the dying of the year; but there is a wonderfully signalled shift from the linear year of autumnal decay to the natural cycle of renewal, during the walk from Uppercross to Winthrop. [...] The reader is reminded that the very purpose of the walk is one of renewal, in Henrietta's reconciliation with her neglected suitor Charles Hayter, and that Wentworth's gesture in helping Anne into the Crofts' gig is a sign of thaw to come (*P* 13–14).

Thus, the reader, like the farmer "meaning to have spring again" (*P* 91), gets a sense of coming renewal, springing from the decaying nature.

Later in the month, the cyclicity is broken by an accident. Like spontaneous Marianne Dashwood, the no less spontaneous Louisa Musgrove falls, or more precisely, deliberately jumps from the Cobb[10] in Lyme.

[10]The OED states that this word can be spelled *cob* or *cobb*. Interestingly, there is another sense to it. In 1884 R. Holland's *Gloss. Words County of Chester* (1886): *Cob*, a blow, generally on the head.

Unluckily and symbolically, considering the final outcome of her "love" for him, Captain Wentworth fails to catch her and the accident is much more serious than Marianne's sprained ankle. Incidentally, this accident is anticipated by the fall and minor injury of little Charles, which prevents Anne from meeting Wentworth at the same time as the other members of the family. This gives her time to psychologically prepare for the unavoidable meeting by hearing her family members speak about him and the impression he made on her two young cousins. In contrast to Marianne and Willoughby whose love story starts with the accident, the flirtation between Louisa and Wentworth ends with it. The fall and, especially, the almost three months necessary for her recovery[11] change Louisa's wishes and inclinations, although they bring a marriage for her anyway: she gets close to Captain Benwick and later marries him. The linearity of the accident can also be seen as a real re-starting point of Wentworth's affection for Anne. From this moment on, their relationship will move onto a different emotional stage: it will enter a different cycle and slowly develop. The accident happens on the border between the land and the sea, out of their familiar home space as well as out of a social season—in an empty seaside resort, which would symbolize their relationship at that moment. We might say that some reckless act was necessary to unblock the situation for Anne and Wentworth, and it could only happen in an unfamiliar liminal space, out of the usual everyday cyclicity of life at home.

During Christmas the Musgroves have their own children, and the Harvilles' children back for the holiday. As in *P&P*, the season is marked by noise, which contrasts with the emptiness of the preceding weeks, when all the family stayed in Lyme because of Louisa's accident, and Lady Russell remarks that she hopes to remember in future "not to call at Uppercross in the Christmas holidays" (*P* 146) expressing her preference for tranquillity and regularity. But there is the irony of the following narrative comment which points up the subjective response to conditions of space: when Lady Russell enters the noisy streets of Bath she does not complain (*P* 146). This period is, however sadly significant to Anne, for these are the last days that she spends in her home space before leaving for Bath. In spatial terms, from this moment on, the action will evolve in Bath and all the important characters will find their way there.

[11] The accident happens in mid-November and Louisa returns home on 2 February (Shapard 488–489).

As in *NA* the two Winter months, January and February, are full of events in *P*. However, unlike Catherine, Anne comes to Bath as to her new home. Her father and sister are eager to construct and furnish their new social space by forming new "honourable" acquaintances and attending fashionable evening events. Anne, though sometimes taking part in these social gatherings with her family, regularly takes time off and creates a parallel space for herself by maintaining a close personal relationship with Lady Russell and visiting her infirm and poor school friend Mrs. Smith. During January, too, Anne's attention is claimed by Mr. Elliot who starts actively courting her. This brief relationship parallels the autumnal flirtations and love stories with "wrong" men in other novels. Anne is hesitant, but though flattered by his attention she does not fall in love with him. And even though Lady Russell imagines them married by Autumn (the time for planned marriages in the novels), Anne does not make such plans. However, the relationship has its importance in helping Anne to recover confidence in her personal attractiveness and thus making her believe in the possibility of attracting again the man she really loves. At the same time, she is conscious that she cannot have any real expectations with regard to Wentworth, for during January she still thinks him bound to Louisa. As in other novels, January is here too a month of waiting. Retrospectively it appears to be the case for Wentworth too: he spends about two months (from mid-December to mid-February) with his brother, after having found out that his supposed engagement to Louisa was considered by people as certain (*P* 263). At this time, he is also starting to realize that he still loves Anne.

As in other novels, February brings news in *P*. The difference here is that this is the last month of the action in the novel. At the beginning of the month Louisa's engagement to Captain Benwick is announced which morally frees Wentworth and unblocks the situation for Anne: "it was not regret which made Anne's heart beat in spite of herself, and brought the colour into her cheeks when she thought of Captain Wentworth unshackled and free" (*P* 182). After the Crofts, Wentworth himself arrives in Bath, followed by the Musgroves. By the end of the month, almost all the characters are together again—the space/time is filled with movement as in Uppercross and Lyme. Strikingly, February events occupy a quarter of the novel (6 out of 24 chapters) and from mid-February on, almost every day is related, thus the reader, at this point, experiences a deceleration of the pace and so senses the approaching resolution. All the cyclical activities, such as walking, visiting, going to a theatre or an evening party, become important elements of a linear chain leading to definitive changes in several characters: Anne and Wentworth, Louisa and Benwick, Mr. Elliot and Mrs. Clay, and later, Mrs. Smith. Suddenly Bath changes its status as a space: from the frightening place of banishment, it becomes a space of a new beginning, a springboard to the different life stage for Anne and Captain Wentworth. The actual engagement of the protagonists takes place in a street and they exchange their first avowals of love only by looking at each other in the company of Charles Musgrove, whose presence is symbolic as Anne had refused him once, because she still loved Wentworth. The cyclicity of the public space frames this linear moment of their lives. Their conversation brings together all the past events that form the chain that has led them to the present situation. Thus, they step over the threshold of the engagement at the very end of February, on the border of Winter and Spring. Significantly, their first meeting and romance eight years earlier took place during Summer and Autumn, thus, they are completing their year of courtship with Spring which starts at the beginning of their second engagement.[12]

Looking at the other years that frame the main action of the novel, I notice that the beginning of *P* differs from the other novels not only by relating linear events but also by conveying the sense of the cyclicity of human life. The first lines of the novel show Sir Walter Elliot consulting the book of his family lineage. The reader is to understand that it is a regular and ritualized activity, for Sir Walter "never took up any book but the Baronetage; there he found occupation for an idle hour, and consolation in a distressed one; there his faculties were roused into admiration and respect, by contemplating the limited remnant of the earliest patents" (*P* 3). Looking at the book with its succession of marriages, births, deaths and redundant first names establishes a cyclical climate to the novel from the very first pages. Moreover, this first chapter also lays out the background of the actual family situation by marking relatively recent linear events as in other novels—the death of Mrs. Elliot 13 years ago (*P* 5) and the entail of the family home—Kellynch-hall—to the closest male relative, as well as a long expected linear event that has still not happened—Elizabeth's marriage, which would be the natural way to assure that Kellynch-hall continues in the family. Moreover, the meeting and the broken engagement between Anne and Wentworth eight years earlier form the base for the

[12] See Shapard *P* timeline.

novel's main action. The past pain continues to influence the present until the happy moment of their second engagement, thus the reader can feel two time lines coexisting throughout the narrative.

The Watsons and *Sanditon*

In *S&S*, *P&P*, *MP*, and *E* the Autumn months are filled with intense social life in more or less large circles, which give opportunities to fall in love whether "truly" or "falsely." Even more strikingly, in her unfinished novel, *The Watsons* (*W*), Austen plunges the reader into this socially exciting time with the opening sentence: "The first winter assembly in the town of D. in Surrey was to be held on Tuesday, October 13th, & it was generally expected to be a very good one; a long list of country families was confidently run over as sure of attending, & sanguine hopes were entertained that the Osbornes themselves would be there" (*W* 79). This is the first ball for the nineteen-year-old Emma Watson, who has just returned home. The acquaintances that Emma makes during the ball—with Tom Musgrave, Mr. Howard, and Lord Osborne—hint at possible emotional entanglements and potentially linear events; unfortunately, of course, they are never realized. In the other novels, this very intense period is also followed by disappointments, which can be more or less definitive. The time is outwardly cyclical. We witness the repetition of ritualized activities: balls, tea or evening parties, walks, shopping, and long days of waiting at home because of the bad weather. However, at the same time, it is full of potential linearity: the new relationships may lead the protagonists to points of no return in a positive or negative way.

The twelve chapters of Austen's unfinished novel *Sanditon* (*S*) take place during a couple of weeks in July. It starts with an accident reminding the reader of similar events in *S&S* and *P*. However, it is towards mid-August that the presumed heroine, Charlotte, sees for the first time the presumed hero, Sidney Parker, thus, the reader could expect to follow the beginning of their love story in the early Autumn.

Sanditon is a seaside town and the main preoccupation is connected with the holiday season: the arrival or non-arrival of new people for a seaside holiday. In her other novels Austen never directly describes any seaside holiday, however, even if in the background, they sometimes have important implications for the development of the story, as with the meeting of Jane and Frank in Weymouth in *E* and the seaside holiday of Isabella's family which occasions the family quarrel. Thus, both occasions bring linear events. Sanditon too seems to be presented as a linear space, where the potential heroines, Charlotte and Clara, are about to experience some life transforming events. Unfortunately, it is not possible to see what place the Summer season would have had in the context of the completed novel, however, this different starting point might indicate a change of structural strategy.

Spatiotemporal Structures: An Overview of the Novels in Their Chronological Order

To close this chapter, I propose to return to the tables presented at the outset and to sketch an overview of the time scheme of each novel in order to highlight how Austen's use of these structuring patterns varies.

As I mentioned, *NA* has the simplest temporal structure, which is equally divided into two main parts: Winter, when action passes in Bath, and Spring, when Catherine stays at Northanger Abbey. Thus, here we have a "one season—one place" scheme. The Winter at Bath represents an open space/time: Catherine is getting acquainted with different possibilities and starts making her own choices. The Spring part symbolically corresponds to Catherine's personal growth. The reader can observe the development of her feelings, relationships, awareness of her own preferences and re-evaluation of her former dreams. The Spring part also has a clear linear direction that leads to the protagonists' marriage. Moreover, the main action of the novel is framed by two chapters: one on pre-history, where the heroine is rapidly growing up, and one chapter on the year after the end, where the last obstacles are resolved and the characters marry.

In *S&S* the temporal structure gets more complex and less equally divided. The first four chapters of the novel precede the main plot and, after a brief presentation of the family history, dwell on the mourning period after Mr. Dashwood's death that lasts during Spring and Summer. These six months allow the necessary time for the psychological recovery of Elinor and, thus, give her the possibility to fall in love. However, with the beginning of the Autumn and the move to Barton the time of the narrative slows down. Austen takes time to set the problematic plot lines that will drive the story towards its resolution. Autumn occupies 40%, that is, almost half of the novel and corresponds to one place—Barton and its environs. The next two parts—Winter and Spring—are more equally divided, each occupying about a quarter of the narrative. Winter corresponds to one place—London, and to the time of emotional gestation and disappointing news. Spring, on the other hand, is more complex in spatial terms. It is a time of a long journey home, from its sad preparation in London, through the critical point at Cleveland, and finally, to the soothing return to Barton. The complexity of places in Spring corresponds to the time of most complex psychological and relational developments. Thus, three seasons are fully present in the novel, with Summer framing the main plot—first as prehistory, and then, as preparation time for Elinor's marriage.

P&P is the least equally balanced novel in terms of seasons. It starts *in medias res* with Autumn, which occupies more than a half of the whole narrative. All the autumnal action happens, moreover, in one county, Hertfordshire: at Longbourn, Meryton, and Netherfield. It is the time when Elizabeth's opinion about Mr. Darcy and Mr. Wickham gets formed, when she is intransigent and confident, supported by her familiar home space. Winter occupies only 5% of the novel, and, after a more animated Christmas period, which effectively "removes" Mr. Wickham as Elizabeth's suitor, it is dedicated to waiting for Spring with its news. Spring and Summer, close in length, are the seasons of movement in spatial and psychological terms, rather like Spring in *S&S*. Elizabeth travels, is confronted to unexpected events, and her opinions as well as feelings gradually change. Noticeably, *P&P* is a novel that has hardly any temporal frame: all the narrative, except the second half of the last chapter, is concentrated on the main action and on one crucial year.

MP, unlike *P&P*, has an important introductory part before the start of the main action. A quarter of the whole novel is given to the preceding years and the preceding Summer. It is taken up with the depiction of the psychological ties between the Bertram family members, Mrs. Norris and Fanny, as well as the inner functioning of the household. Moreover, during Summer, the Crawfords are carefully "inserted" into the picture and their impact on other characters thoroughly illustrated. The rest of the year is more or less equally divided between the remaining three seasons, with Winter—the most distressing time for Fanny—occupying the biggest part. Mansfield Park, as even the title of the novel reveals, is the principal place during all seasons; even when Fanny is away in Portsmouth, the reader has an opportunity to experience the Spring at Mansfield through Fanny's mental projection discussed earlier.

E is composed of four fairly equally balanced seasons, though Winter slightly dominates the rest. Autumn is centred on Emma's first unlucky attempt at match making with Mr. Elton, though its full meaning is only revealed in Winter—on Christmas Eve. After the reflective and, thus, eventless January, we have a very active and detailed February which brings a major change to the plot with the arrivals of Jane and Frank. This period also corresponds to Emma's second serious misjudgement of the situation. Spring is a period "in between," full of hesitations and suspended physical and emotional activity. It is also slightly shorter than the other seasons. Finally, Summer is the time of resolution, disclosure and self-understanding. Noticeably, here, unlike *MP* with its "transplanted" heroine Fanny, the protagonist—Emma—is identified with the main place—Highbury—where all the action of the novel happens.

P differs from the other novels: even though externally Austen goes back to a shorter time line, as in *NA*, it is internally thicker. The past years are constantly present during the time of the main plot: first of all, Anne's broken engagement with Wentworth, but also her mother's death that destabilized their family. The past also haunts—directly or indirectly—other characters: Mrs. Smith, Mr. Elliot, Mrs. Musgrove, and Lady Russell. In terms of seasons, Summer is the shortest in the narrative, although it is aligned with the Summer eight years earlier. It is also the season of resignation to the approaching departure from Kellynch. Autumn and Winter are equally divided, respectively corresponding to the time spent at Uppercross and Bath. Autumn is the time of false impressions, accidents and recovery that leads to a constructive and linear Winter.

Austen thus goes from a simple binomial structure in *NA*, which makes relatively slight use of the seasonal pattern in relation to the characters to a more complex seasonal architecture that comprehends the main plot in one year in *S&S*, *P&P*, *MP*, and *E*. Moreover, in *MP* and *E* Austen makes the fullest use of the possibilities of meaning that these seasonal structures offer. Her last novel *P*, however, though containing only three seasons during the time of the main action, offers a thicker, two-layered time line, where the past is addressed and repaired by the present.

Works Cited

Primary Sources

Austen, Jane. *Northanger Abbey*. Ed. Barbara M. Benedict and Deidre Le Faye. Cambridge: Cambridge University Press, 2006a.

Austen, Jane. *Sense and Sensibility*. Ed. Edward Copeland. Cambridge: Cambridge University Press, 2006b.

Austen, Jane. *Pride and Prejudice*. Ed. Pat Rogers. Cambridge: Cambridge University Press, 2006c.

Austen, Jane. *Mansfield Park*. Ed. John Wiltshire. Cambridge: Cambridge University Press, 2005a.

Austen, Jane. *Emma*. Ed. Richard Cronin and Dorothy McMillan. Cambridge: Cambridge University Press, 2005b.

Austen, Jane. *Persuasion*. Ed. Janet Todd and Antje Blank. Cambridge: Cambridge University Press, 2006d.

Austen, Jane. *Later Manuscripts*. Ed. Janet Todd and Linda Bree. Cambridge: Cambridge University Press, 2008.

Secondary Sources

Beveridge, William H. "Marriage and Birth Seasons." *Economica,* New Series, 3.10 (1936): 133–161.

Bree, Linda. Introduction. *Persuasion*. By Jane Austen. Ed. L. Bree. Ontario: Broadview literary texts, 1998. 7–37.

Davidoff, Leonore. *The Best Circles. Society, Etiquette and The Season*. London: Croom Helm, 1973.

Deresiewicz, William. *Jane Austen and the Romantic Poets*. New York: Columbia University Press, 2004.

Edgecombe, Rodney, S. "Change and Fixity in *Sense and Sensibility*" in *Studies in English Literature, 1500–1900,* Vol. 41, No. 3, Summer 2001, pp. 605–622.

Hutton, Ronald. "Modern Pagan Festivals: A Study in the Nature of Tradition." in *Folklore* 119.3 (2008): 251–273.

Lotman, Yuri M. *Universe of the Mind*. Trans. Ann Shukman. Bloomington and Indianapolis: Indiana University Press, 1990.

Lynch, Deidre Shauna. "'Young ladies are delicate plants': Jane Austen and Greenhouse Romanticism." *English Literary History* 77.3 (2010): 689–729.

Mullan, John. *What Matters in Jane Austen?* London: Bloomsbury, 2012.

Pool, Daniel. *What Jane Austen Ate and Charles Dickens Knew*. New York: Touchstone, 1993.

Shapard, David M., ed. *The Annotated Persuasion. Jane Austen*. New York: Anchor Books, 2010.

Todd, Janet, ed. *Jane Austen in context*. Cambridge: Cambridge University Press, 2005.

CHAPTER 3

Changing Homes: Transformations in Space and Time

Austen's novels are structured around permanent or temporary home spaces. During the narrative, characters move from one life space to another—passages which are sometimes marked by various thresholds and connecting spaces such as windows and doors—and these trajectories are related to social and psychological transformations. According to F. Saggini and A. E. Soccio, both words, "house" and "home," are connected to a shelter that satisfies humans' "primary needs," but the word "home" reflects intimacy and the inhabitant's values, while "house" stands more for the building itself (2). Moreover, they acknowledge that:

> In considering such a space, one cannot avoid confrontation with crucial themes often re-elaborated in literature and culture, representing the house/home as the metaphor for either inner, psychological space, or entire universe, on the one hand, and the ordering and spatial representation of an artist's *Weltanschauung* on the other. "For our house", says Bachelard, "is our corner of the world."[1] (2)

The image of home in Austen's novels often implies affective bonds. Therefore, in this chapter, I examine the characters' psychological relationship to different houses/homes and its evolution through time. I also look at the characters' place in the different households, not forgetting the

[1] Bachelard, 4.

R. Baublytė Kaufmann, *The Architecture of Space-Time in the Novels of Jane Austen*, https://doi.org/10.1007/978-3-319-90011-7_3

social aspect of these relationships. Overall, there are four types of homes: childhood/original homes, temporary homes, dream homes, and independent adult homes, which in some cases overlap. These different types of homes reveal different psychological phases of the characters that change with time: most of them grow out of their childhood homes and generate the images of their dream homes in their heads while lingering in temporary home spaces, until they can, finally, found their mature adult homes. Tellingly, the original, temporary, and, most of the time, adult homes are conceived as a set of relations rather than as physical buildings, and so they are not described in detail. The domains like Pemberley, Rosings, Donwell Abbey, Northanger Abbey, and Mansfield Park are, on the other hand, thoroughly depicted, when they are seen, at least by some of the characters, from the dream house/home perspective and when they are not (yet) inhabited by these characters.

All six novels, as well as the two unfinished ones, start with a situation of disequilibrium at the heroines' original/childhood homes. Catherine Morland in *NA* is dreaming of adventures that her home space cannot provide. Similarly, Emma in *E* is looking for new projects to animate her life in the "deserted" house after Miss Taylor's marriage. Elizabeth Bennet in *P&P* has outgrown her original, though entailed, home and is ready to quit it for good, whereas the Dashwood sisters in *S&S*, little Fanny Price in *MP*, and Anne Elliot in *P* are forced to move out of their childhood home spaces. Generally, childhood home experiences condition the heroines' future trajectories in psychological and social ways.

There are two types of temporary homes in the novels. Most frequently we see the places that characters visit and where they stay for several weeks with or without forming emotional attachments to them. In some other cases, however, their original home (or its substitute) changes status, due to external circumstances, and thus, becomes a home that the characters will have to leave soon. Sometimes, these homes are not originally theirs, but they function as if they were. Moreover, some important linear events in the novels happen in temporary spaces outside the heroines' reach.

The lengthy stays in somebody else's home or in rented homes often correspond to the London or Bath season. We get the most detailed picture of the season in London from *S&S*, as I will discuss later. In *MP*, all the young characters, except Fanny, go to London during the season, but we find out what is going on there only through Edmund's and Mary Crawford's letters, which Fanny gets while staying in Portsmouth. These letters convey the development of Edmund's feelings for Mary (*MP* 488–489), Mary's hopes concerning Edmund and Henry (*MP* 481–482), and, later, Tom's illness (*MP* 494) as well as Maria's elopement (*MP* 506). Town is a space where important and troublesome events take place during Fanny's absence, and it thus contrasts with her eventless and isolated first home space. There is also a contrast in terms of time: while Fanny suffers from solitude and boredom during these seemingly endless weeks in Portsmouth, the same weeks in London are packed with events. On the other hand, Fanny judges London a corrupt space: she is "disposed to think the influence of London very much at war with all respectable attachments" (*MP* 501). In general, even though we do not see the real temporary homes that characters inhabit in London and their relationship to them, the time they pass there brings important transformations. Thus, one of the pivotal points in the novel, Mr. Crawford and Maria's elopement, happens out of the heroine's and reader's sight in a space that is unfamiliar to her.

There are other instances of such spatially concealed shifting points in the novels, which do not necessarily happen during the season. In *NA*, General Tilney's decision to turn Catherine out of his home is made during his week in London (*NA* 227) away from his children and their guest. However, the gossip that he hears in some unnamed house speeds up the resolution of Catherine and Henry's relationship. Similarly, in *P&P*, Lydia's sudden "love story" with Wickham and their elopement, which have such serious consequences for Elizabeth and Mr. Darcy, take place in Brighton (where the regiment is settled for the Summer) which is never described in the novel. Similarly too, in *E*, the manipulative Mrs. Churchill suddenly dies somewhere in Richmond and removes the main obstacle for Frank and Jane's happy future together. Thus, it appears that the contingent events that are not directly dependant on the main characters' actions, take place in some strange, unfamiliar space, shielded from the main characters' and the reader's eyes. Symbolically, such spaces would represent the uncontrollable and unpredictable side of life—hidden surprises—that we are all exposed to.

The notion of dream home opposes that of original/childhood home. Normally, it represents a future space, although, as I will show later, it can also epitomize a lost past space. When talking about the phenomenon of a dream house, Bachelard explains that:

> This dream house may be merely a dream of ownership, the embodiment of everything that is considered convenient, comfortable, healthy, sound, desirable, by other people. It must therefore satisfy both pride and reason, two irreconcilable terms. […] Thus the dream house must possess every virtue. However spacious, it must also be a cottage, a dove-cote, a nest, a chrysalis. (61, 65)

For Bachelard, the space we dream of is complex, for we do not always dream of the same type of house and, thus, we have "our cottage moments and our palace moments" (63). Basing myself on this, I consider the image of the dream home in the novels, as a space that embodies all the most desirable qualities from the character's point of view and/or excites her imagination.

Each of Austen's novels has its dream home/house, though not all of them are equally important for the development of the plot. In general, the dream houses are connected to the heroines' future or past, so, while some of them manage to transform these spaces into their adult homes, others have to renounce them and find home elsewhere. Catherine Morland's dream house in *NA* loses its emotional status and the heroines of *S&S* and *P* physically lose their dream homes, whereas the heroines of *P&P*, *MP*, and *E* become the mistresses of their dream homes. According to Purdie, the "perfect" houses are the final reward for protagonists who "after proving themselves through various tests and tribulations, attain a position allowing them to become possessors of their own homes (whether directly or through marriages)" (17). However, even though all the heroines have to pass through trying psychological "tribulations" and face phases of disillusionment and repeated self-evaluations, not all of them have the same expectations for their future: Marianne's "poverty," for example, is Elinor's wealth (*S&S* 106). Thus, the home-rewards they get, logically, also vary.

At the end of the novels some of the dream homes—Pemberley, Mansfield Park and, partly, Hartfield and Donwell Abbey—are transformed into adult homes for the heroines and their husbands, while other characters found their adult homes without having created any prior elevated image of them. Thus, adult homes may contrast with dream homes, as reality contrasts with imagination. On the other hand, an independent adult home does not have a prior connection with the maternal-cradle aspect of the original home; it is a home founded by choice and everything is to be constructed in order to transform it into the heroine's own home. In this way it differs from a trans-generational domain, such as Pemberley, where the new heirs have to fit themselves into the ancestral cyclicity.

The life trajectories sometimes oppose an individual to her/his family: certain characters, especially the heroines, do not repeat the ancestral cycle and do not fit within the borders of their family history, but push into newly transformed space thus accomplishing a linear change. This is, for example, the case with Elizabeth Bennet and Fanny Price marrying into much greater houses and richer families than their original ones. In general, both types of trajectories—family and individual—are articulated in relation to houses/homes and the time passed in them or away from them.

In this chapter I will examine the four types of homes in each novel separately. I will finish with a section on the connecting spaces—doors and windows—and compare their functions in the novels.

Northanger Abbey

NA is one of two novels (the other is *MP*) where the reader gets a glimpse of the heroine's childhood. However, Catherine's childhood house is not actually described in the beginning of the novel. The place where she grows up is rather an outside space: "she was moreover noisy and wild, hated confinement and cleanliness, and loved nothing so well in the world as rolling down the green slope at the back of the house" (*NA* 7). The image of a slope behind the house presupposes a protected, secure home, traditionally "built in a hole" as Mr. Parker from *S* would say (*S* 156), and even though Catherine is said to hate confinement, she amuses herself in the outskirts of the house by rolling down the slope towards it. This physical dynamism and freedom from limits that we gather from the description will relocate from Catherine's body to her mind when she will grow up and form her wild imagination.

Gender, however, might be important in this case for the female education of the time was aimed at "instilling Decorum, Manners and Deportment" (Le Faye, 89) and not at strengthening girls' bodies by physical exercise. Bearing that in mind, we can see that Austen presents young Catherine as a very unusual girl (*NA* 6) with indulgent parents who did not force her to acquire the typical female accomplishments of the time. However, in teenage years her physical energy takes a more socially acceptable form and this transformation is positively evaluated by her parents: "her love of dirt gave way to inclination to finery, and she grew clean as she grew smart; she had now the pleasure of sometimes hearing her father and mother remark on her personal improvement" (*NA* 7).

Alongside this, the reader gets a notion of Catherine's family as a happy household, where things evolve smoothly without early deaths, entailments, or any other of the traumatic changes that happen in the other five novels. Her big and happy family seems to take the place of the house itself and plays the role of refuge when Catherine needs it at the end of the novel. Most of the narrative takes place out of Fullerton. Catherine only returns after she is driven out of Northanger Abbey. Her coming home is presented not so much as a return to a geographical place but as a return to her family circle where she feels loved and protected: "in the embrace of each, as she stepped from the carriage, she found herself soothed beyond any thing that she had believed possible" (*NA* 241). The image of an embrace conveys a sense of roundness and security, which is a metonym for the stability of her childhood home.

Unlike Fullerton, Northanger Abbey—a childhood home for Frederick, Henry, and Eleanor Tilney—is an unsatisfactory and oppressive home space, which generates a wish to escape. Their household, especially after the death of their mother (*NA* 184) is not a happy one: the tension between the inclinations of the children and the firm, oppressive authority of their father (*NA* 191) does not make it a comfortable place to live. It does not even convey the impression of being a home. The house functions according to General Tilney's strict rules and thus resembles a school or other public institution, or, paradoxically, a real abbey with its regulations. However, as an unmarried woman, Eleanor is the only member of the family obliged to stay in the Abbey. Frederick, the heir, is also occupied in the army, and Henry, a clergyman, has his parish living, where he could pass all his time, if he did not think of his lonely sister (*NA* 184). Being often alone in this house (*NA* 160) Eleanor clings to the image of her departed mother, and, at the same time, of her happy childhood, by inhabiting her mother's favourite places, walking in the grove she liked and giving into "the delightful melancholy which such a grove inspired" (*NA* 184). Thus, Eleanor creates a certain space of comfort, based on past experiences, within a hostile and restrictive home. On the other hand, like Anne Elliot, who was persuaded into breaking her engagement with a poor and untitled sailor, Eleanor is not allowed to marry whom she wants, which complicates her only possibility of escape from her original home. Fortunately for her, the novel ends with her lover's "unexpected accession to title and fortune" (*NA* 260) and "her removal from all the evils of such a home as Northanger" (*NA* 260).

The occasions to stay in temporary homes appear when Catherine goes to Bath during the season. She stays there for six weeks (*NA* 10) with the Allens during January and February and meets other families that are there for the same purpose: the Thorpes and the Tilneys. We learn that Catherine stays in Pulteney-street (*NA* 46), that she calls the lodgings "home" (*NA* 56) and that she dreams of Henry Tilney in her room (*NA* 22). However, the most important places in Bath are the public ones. The main meetings and conversations take place in the Pump-room, the Lower-rooms, the Upper-rooms, and the theatre. Thus, the temporary lodgings in Bath only provide a place to sleep and dress, and to leave to mix with society as soon as possible. The only house that hosts an event is the Tilneys' rented house in Milsom-street where Catherine is invited for a spiritless and psychologically uncomfortable dinner (*NA* 131).

Northanger Abbey, where Catherine goes after Bath, represents a dream home space. We find a dream house in the most Bachelardian sense—an imaginary space that has little to do with the real Northanger Abbey, as it is depicted in the novel. Catherine Morland "builds" her fantasy house inspired by gothic literature, which is doomed to collapse once confronted with reality. Northanger Abbey, like all dream houses, is also a linear space, a goal where the heroine goes and is psychologically transformed.

The moment when Catherine is asked to accompany Eleanor Tilney to her home is presented as emotionally overwhelming: "Northanger Abbey!—These were thrilling words, and wound up Catherine's feelings to the highest point of extasy" (*NA* 142). The image of an old abbey-house in her mind embodies all her dreams of gothic adventures. Moreover, these fantasies are intensified by the prospect of staying close to Henry and possibly having her feelings reciprocated. The event provokes in Catherine a blissful state of jumbled day-dreaming:

> She was to be their chosen visitor, she was to be for weeks under the same roof with the person whose society she most prized—and, in addition to all the rest, this roof was to be the roof of an abbey!—Her passion for ancient edifices was next in degree to her passion for Henry Tilney—and castles and abbies made usually the charm of those reveries which his image did not fill. (*NA* 143)

The image of Henry melts into the images of his father's "legendary" house, one strengthening and extending the other in her imagination. The image of "roof" is equivocal: on the one hand, it covers, shelters her desire to live with the man she admires in the space she dreams about. However, at the same time a "roof" is limiting and obstructing, which may symbolize her incapacity to see the real abbey as it is. Catherine's dream house is in the form of an abbey and not "hall, place, park, court and cottage" (*NA* 143). Bringing together the passages from the gothic stories she has read this elates her expectations to an unreasonable level. Eleanor's sadness about her lonely and psychologically restrained life there, Henry's irony, the General's severe household rules and Frederic's avoidance of the place do not immediately awaken her to reality.

Noticeably, Catherine thinks of Northanger Abbey in terms of a dream house, but not in terms of a dream home. She concentrates on architectural features and the adventurous stories that should go with such surroundings. However, she does not imagine living there with Henry Tilney, or visiting once she is married and living elsewhere (Henry is not the heir of the Abbey), as the heroines of other novels do when thinking of their dream houses. We see that after her arrival at the Abbey, Catherine observes every architectural detail trying to match it to the house of her imagination, which is composed from features she has read about, and, noticing the discrepancies, she starts to feel disappointment:

> The furniture was in all the profusion and elegance of modern taste. [...] The windows, to which she looked with peculiar dependence, from having heard the General talk of his preserving them in their Gothic form with reverential care, were yet less what her fancy had portrayed. [...] To an imagination which had hoped for the smallest divisions, and the heaviest stone-work, for painted glass, dirt and cobwebs, the difference was very distressing. (*NA* 165–166)

The readers see a real, modernized house, which is there to form a representational home that affirms the Tilneys' social status,[2] and, at the same time, feel Catherine's measuring it against her imagined dream house. However, Catherine does not easily renounce her fantasy world. She "fights" with an "old-fashioned black cabinet" (*NA* 172) and its drawers during the stormy night in her room, and tries to resolve the "mystery" of Mrs. Tilney's death by imagining a domestic murder (*NA* 192). Unfortunately for her, she is forced to face the banality of an old laundry list (*NA* 177), as well as the implacably sober Henry Tilney's explanations concerning his mother's physical death (*NA* 203).

This rational conversation with Henry, and her fear of his seeing her fantasies as absurd and insulting, put an end to her "visions of romance" (*NA* 204). Thus, Catherine's image of the Abbey as a dream house ends in disenchantment and this happens before she sees her future adult home—Woodston Parsonage:

[2] Mirella Billi sees this description of Northanger Abbey as a representation of General Tilney's inhospitable character "for whom the house is only a status symbol [...] meant to impress the visitors" (230). I'm concentrating rather on Catherine's point of view, and the image of the abbey in her mind.

> the abbey in itself was no more to her now than any other house. The painful remembrance of the folly it had helped to nourish and perfect was the only emotion which could spring from a consideration of the building. What a revolution in her ideas! [...] Now, there was nothing so charming to her imagination as the unpretending comfort of a well-connected parsonage, something like Fullerton, but better. (*NA* 218)

The passage shows Catherine is cured of her gothic fantasies and, as Johnston puts it, she starts wishing for "a more concise and comprehensible residence than the Gothic, which is fraught with misleading objects" (*NA* 198). The process advances with their visit to Henry's parsonage and Catherine's emotional acceptance of this space (*NA* 220) that is later to become her home. Gothic images of an abbey in her head are supplanted by the prettiness of a cottage "among the trees" and "the windows reaching to the ground" with a view over meadows (*NA* 220). Woodston is presented as an idyllic and pastoral space with its "puppies just able to roll about." young shrubs, meadows, and fresh air (*NA* 220–221). The house itself is seen through the enthusiastic eyes of Catherine who appreciates its "honest simplicity" and prettiness (*NA* 220). In Woodston she finds her new dream home, much closer to her authentic self than the image of a gothic abbey.

Noticeably, Catherine begins to appreciate something that reminds her of her childhood home space, even though her imagination is still busy at work by transforming her original home image into something "better." Her excursion to the adventurous world of wild fantasies is over; she starts to appropriate the familiar cyclicity of living in a parsonage, though, preferably, in an "improved" one. Thus, for her adult home, Catherine finds an equivalent space to her childhood house: she does not really make a linear step in social or psychological sense, but learns to recognize her "comfort zone" and stays within it.

Sense and Sensibility

In Austen's second novel *S&S* the leaving of the childhood home is much more painful, even dramatic than in *NA*. Although the opening chapter of the novel depicts the Dashwoods and their old uncle as a happy family (*S&S* 3), the deaths of the old gentleman and then of Henry Dashwood overturn the situation. Thus, the main action of the novel starts when the Dashwood girls legally lose their home by being "degraded to the condition of visitors" (*S&S* 9). This way, the cycle of inheritance is cut, and the family of four females becomes temporarily homeless. During the next several months the girls and their mother go through the process of mourning and their former home gradually transforms itself from an intimate to a strange space. The reader can observe this process from the depiction of Mrs. Dashwood's state of mind in Chap. 3:

> Mrs. Dashwood remained at Norland several months; not from any disinclination to move when the sight of every well known spot ceased to raise the violent emotion which it produced for a while; for when her spirits began to revive, and her mind became capable of some other exertion than that of heightening its affliction by melancholy remembrances, she was impatient to be gone, and indefatigable in her inquiries for a suitable dwelling in the neighbourhood of Norland; for to remove far from that beloved spot was impossible (*S&S* 16).

The extract shows that the notion of home does not only apply to a house and its grounds. It has a second circle that englobes the region in which the house is situated. Thus, once the first stage of mourning is done and the separation from the house itself connected to all the memories of happily living there becomes less painful to imagine, there is still a second stage of mourning to complete. Staying in the same area seems like keeping a part of the original emotional home and, at the same time, a part of that happy past. Moving away to some rented house, unconnected with the family history, means moving to the unknown, geographically, socially, and psychologically, and losing the familiar background in literal and figurative ways. Moreover, the death of Mr. Dashwood condemns his widow and daughters to descend in their social standing. Mrs. Dashwood seems to need more time than her eldest daughter to comprehend this, for she is said at first to look for a large and comfortable house such as she used to have (*S&S* 16), and it takes another emotionally trying event to complete her disillusionment.

The final decision of Mrs. Dashwood to quit the area of Norland comes as an emotional response to her daughter-in-law's hostility. Consequently, she declares that "to remove for ever from that beloved place would be less painful than to inhabit or visit it while such a woman was its mistress" (*S&S* 27). We can see that Mrs. Dashwood perceives her home not as a concrete life space, but rather as an emotionally comfortable space. Once this comfort is lost for good she can detach herself without much hesitation and move on. Marianne Dashwood, like her mother, demonstrates a very strong emotional attachment to her former home. However, in her soliloquy before quitting Norland she refers to it as this particular house that stands in this particular space and forms an unmovable symbolic image of her childhood home:

> "Dear, dear Norland!" said Marianne, as she wandered alone before the house, on the last evening of their being there; "when shall I cease to regret you!—when learn to feel a home elsewhere!—Oh! Happy house, could you know what I suffer in now viewing you from this spot, from whence perhaps I may view you no more!" (*S&S* 32)

Marianne is conscious that she is leaving her childhood home together with her childhood years and that she still hasn't finished mourning her past. The farewell is painful, for she thinks that she will not "feel a home" somewhere else "naturally" as she feels in her childhood space, but she will have to consciously learn it. The contrasting dramatic words used in this speech also reflect Marianne's impulsive and passionate character: the house is "dear" and "happy," while she is regretting and suffering. At this moment of leaving, Norland becomes a liminal space for the Dashwood women. It represents a threshold to a different life: all that is out of its boundaries is an unknown outside space including that far away cottage in the unfamiliar county that they have rented without having ever seen.

However, after settling at Barton, it still takes some time for the heroines to emotionally transfer their home space from Norland. About two weeks after their arrival the Dashwoods feel that: "the house and the garden, with all the objects surrounding them, were now become familiar, and the ordinary pursuits which had given to Norland half its charms, were engaged in again with far greater enjoyment than Norland had been able to afford, since the loss of their father" (*S&S* 48). Here Austen rationally proposes the idea that home is the place where one feels comfortable and actively participates in everyday activities. She describes the process of settling in by showing the girls "placing around them their books and other possessions, to form themselves at home" (*S&S* 35). The unpacking of Marianne's pianoforte and fixing of Elinor's drawings to the walls (*S&S* 35) also help to "personalize" the house. The phase of construction, thus, replaces the period of loss and mourning.

Unlike the Dashwoods who are evicted from their home, Edward Ferrars feels estranged from his original home, which seems to be in London with his mother (*S&S* 407; the Norfolk estate he was supposed to inherit is never referred to as their country home). However, unlike Henry

Tilney, who has his own parsonage, once Edward "is dismissed for ever from his mother's notice" (*S&S* 304) after the discovery of his secret engagement, he seems to have nowhere to go but stays in lodgings and taverns (*S&S* 304). Edward's situation is exceptional in the novels, for normally the eldest sons of the wealthy families do not have to worry about financial matters. Through the narrative Edward wanders from one place to another: he stays in Norland with his sister, then goes to visit his old school master and his secret fiancée, afterwards he visits the Dashwood girls and finally goes back to his mother in London, but he never talks of home. Arguably, the disagreements with his mother concerning his career and marriage (*S&S* 18) make him wish to spend most of his time elsewhere (*S&S* 17) which, gradually, creates an estrangement from his childhood home. Moreover, he is unhappy about his engagement to Lucy Steele but does not allow himself to break it.

In a way, Edward's life is suspended in a transitory state: he is living according to a social cycle of long- and short-term visits to his family and friends, because he does not see any possibility of making a satisfactory linear decision concerning his future. Moreover, he has no home space that would serve as a psychological refuge—wherever he goes he is confronted with expectations that he cannot meet: his mother and his sister are dreaming of his political career and advantageous marriage, Lucy Steele is waiting for an advancement in their engagement, while Elinor Dashwood is expecting a proposal (at least before she finds out the truth from Lucy). Thus, by being unable to make any radical decision and give his life a clear direction, Edward relies on time to bring transformations.

Like in *NA*, an invitation from Mrs. Jennings to London during the season provides the reader with a different example of a temporary home. The difference lies in the type of house that the Miss Dashwoods are staying in: it is not just an anonymous rented house, but Mrs. Jennings' London home, her "settled habitation" (*S&S* 174) where she goes after having spent "a large portion of the year at the houses of her children and friends" (*S&S* 174) and which carries her social stamp. Moreover, as both girls are under the age of 21 (Elinor is 19 and Marianne is 16), Mrs. Jennings officially becomes their temporary guardian and moral guide during the time of the visit. The Dashwood girls stay for more than two months in this place and together with their host, they form a temporary household with its own functioning habits. Fortunately, Mrs. Jennings, though sometimes irksome, is maternal enough to create a home-like refuge for the girls during this emotionally trying Winter. It is important that the Miss Dashwoods share the former Charlotte Parker's bedroom (*S&S* 182) during their stay. Sleeping in the same room affects their emotional relationship for they are exposed to each other's intimate space/time. Unlike at Barton, where Marianne could shut herself in her room to cry and leave her mother and sister wondering downstairs about the state of affairs with Willoughby (*S&S* 87, 95), here her emotional crisis is completely exposed to Elinor's notice, reaction, and intervention, because their bedroom is the only intimate space that they have in Mrs. Jennings' house. Thus, in this shared bedroom they have their most intimate and revealing conversation about Marianne's relationship to Willoughby and, as Mullan comments, "it is difficult to imagine the conversation taking place anywhere but a bedroom" (28).

Unlike Bath in *NA*, a significant part of the novel's action goes on inside, in private spaces: ladies are often at home receiving visitors; if not, they are invited to dinners at houses of their extended family, and once, to a private ball, where Willoughby jilts Marianne. After this event, the London house is there to contain and shield Marianne's powerful emotions. Sometimes it functions as a reassuring hiding place, for under the pretext of indisposition she can stay away from visitors in her room, and sometimes as a prison:

> The restless state of Marianne's mind not only prevented her from remaining in the room a moment after she was dressed, but requiring at once solitude and continual change of place, made her wander about the house till breakfast time, avoiding the sight of every body. (*S&S* 206)

Not being able to wander alone outside and not having the possibility to isolate herself completely from other people in this house Marianne starts wishing to go home, to Barton (*S&S* 217). It is at this moment that the cottage inhabited by her loving mother acquires the most profound status of a home space. The distance, the absence, and the happiness that she has experienced there, contrasting with the unhappiness in London, transform Barton from a substitute home to the real home. Moreover, bearing in mind that Marianne does not yet believe in "second attachments," she must consider the cottage as her permanent home. Instinctively, she seeks the remoteness of Barton in order to survive her emotional storm and the painful curiosity of the surrounding social circle:

> Well then, another day or two, perhaps; but I cannot stay here long, I cannot stay to endure the questions and remarks of all these people. The Middletons and Palmers—how am I to bear their pity? (*S&S* 217)

On the other hand, returning to the house where her mother lives would be like returning to the secure childhood space. For Bachelard, a house-home has got "maternal features," it is perceived like "a large cradle" (7), and we can strongly feel this conflation of home and mother here. However, in Marianne's case, she is dreaming of a second home and not her original home from which she has been banished, as I discussed earlier. The image of Barton cottage as her real home space, where her mother can take care of her, becomes more deeply anchored with every exterior challenge that she has to face.

Before bringing her back home to Barton, Austen makes her pass through another temporary space—the Palmer's house—as well as through another temporary state—her illness. Cleveland is presented as an average large country house "like every other place of the same degree of importance" (*S&S* 342), and Marianne appreciates the "country liberty of wandering from place to place in free and luxurious solitude" (*S&S* 343). However, it also represents an additional emotional trial for her as Cleveland is close to Willoughby's estate:

> Marianne entered the house with an heart swelling with emotion from the consciousness of being only eighty miles from Barton, and not thirty from Combe Magna; and before she had been five minutes within the walls [...] she quitted it again, [...] her eye, wandering over a wide tract of country to the south-east, could fondly rest on the farthest ridge of hills in the horizon, and fancy that from their summits Combe Magna might be seen. (*S&S* 343)

From this passage we can see how Marianne tends to go beyond physical and rational limits, how her imagination, coupled with her desires, carries her away from reality, and, away from her body. By means of an illness her body calls her back "inside": through the physical pain and the confinement within these very walls, she learns her lesson of "sense." This episode is marked by images of intermediateness and temporality: she is in a strange but emotionally charged place mid-way to home, suspended between life and death and in a psychological place of fragility.

In *S&S* there are two dream homes: the Allenham house and Combe Magna, which are, however, much less elaborately depicted and significant to the plot than Northanger Abbey. Both houses are connected to

Willoughby, as they are his presumed future home and his actual home and both of them excite Marianne's imagination. The Dashwoods discover the existence of Allenham before meeting Willoughby. They are attracted to the mansion situated "along the narrow winding valley" (*S&S* 48) because it reminds them of Norland, their lost childhood home and, thus, "interested their imagination and made them wish to be better acquainted with it" (*S&S* 48). Therefore, we can deduce that in their case the imaginary dream home would be equivalent to their lost Norland in its sublimated aspect. In other words, by attributing to an unfamiliar, though externally similar, house the best qualities of their former home they transform the pain of losing their childhood home into positive emotions and start dreaming of the future. Finding out that Willoughby was to become the heir adds another dimension to their interest: believing in his forthcoming union with Marianne they can look at it as at her probable future home. Marianne and Willoughby's "secret" visit to Allenham strengthens the general expectation of their engagement, and, moreover, gives Marianne a tangible location to set her dreams in—she describes the house to Elinor "with delight" and even talks about plans for its improvement (*S&S* 81).

Unlike Allenham, Combe Magna is a place that Marianne never sees during the novel. However, she dreams about it and imagines distinguishing its roof over the horizon (*S&S* 343) when she is wandering alone in Cleveland, right before her nearly fatal illness, as I discussed earlier. The house embodies the image of the loving Willoughby that she has already lost at that point in the narrative. It is in this house that her present from him—a horse bred on his estate—is still "waiting" for her (*S&S* 68), as if to keep their story alive.

Elinor and Edward create the first new adult home of the narrative. Tellingly, it is a parsonage—a life-long, though not inheritable, home space. Unlike Catherine Morland, who has to wait passively while Henry is "improving" their home (though paying attention to her desires expressed during her visit to Woodston, [*NA* 259]), Elinor and Edward cannot bear to wait "till everything was ready" (*S&S* 424) and take an active part in the construction of their house, directing everything according to their own wishes (*S&S* 425) and, thus, making it really theirs. The creation of their new home space is completed once the visitors come, which means that from this moment this home becomes part of a social cyclicity and starts functioning as an independent household.

Marianne, unlike her sister, and more "romantically," enters an established and inheritable Delaford House after her marriage to Colonel Brandon: "she found herself at nineteen, submitting to new attachments, entering on new duties, placed in a new home, a wife, the mistress of a family, and the patroness of a village" (*S&S* 429–430). Thus, Marianne is said to become an adult in social terms. Moreover, without giving any details about the actual house or grounds and their life there, the author conveys the impression of the couple's transformed states of mind that the foundation of their common adult household must have reinforced: Marianne gradually learns to love Colonel Brandon, while he gets back his original "cheerfulness" (*S&S* 430). By living together and sharing household duties they develop their relationship and contribute to each other's psychological well-being. Their union resembles a therapy, for they manage to cure one another from the emotional traumas of the past.

Pride and Prejudice

While the Dashwoods have to leave their original home after the death of their father at the beginning of the narrative, the Bennet girls in *P&P* are facing the same danger in the future for their father's estate is entailed. Thus, Longbourn, their original home, has, in a way, become a temporary home from the moment that Mr. and Mrs. Bennet realize that they will not have a son. However, unlike the previous novels, a significant part of *P&P*'s action takes place in or around Longbourn and the principal issue for the girls during the time of the narrative is to quit their childhood home by marrying. On the other hand, this home is represented more by the Bennet family and its ways of functioning than by the house and/or the area itself, though we do see things happening in their drawing room and the library, as well as in the shrubbery and in the "prettyish kind of a little wilderness on one side of the lawn" (*P&P* 391). As in *NA*, the members of the family fill up the home space and the only person who explicitly refers to the Longbourn house is Mrs. Bennet who regularly anticipates the regret of losing it at her husband's death.

The Longbourn household is a very conflictual space. The motor of its tensions is Mrs. Bennet who has acquired this house by marrying a man of a higher social status than herself, as we first learn in chapter 7, thus breaking through her family cycle, and who is very anxious to maintain it for her children or to secure them another equally or, preferably, more eligible home. For her the size and type of the house equals social status, which, she feels, is the central pillar in life.[3] Mr. Bennet's sarcasm, passivity, and

[3] Mrs. Bennet's view reflects eighteenth-century thinking about the importance of home space "where the house represents the wealth, class and occupation of its owner" (Varey, 19).

disinvestment provoke her and generate the conflictual atmosphere in which their daughters grow up. However, the different parts of the house and its grounds are not touched in equal degrees by these domestic storms. Most of the verbal matches between Mrs. and Mr. Bennet go on in the sitting room when all the family is present, as in chapter 2 (Vol. I), where Mr. Bennet is teasing his wife by pretending that he did not visit Mr. Bingley. Symbolically, the sitting room represents the very centre of the household and the place where the family members come together. Thus, the Bennets don't have a strong, indivisible nucleus that would work as a refuge in times of psychological need as we have seen in *NA* or between the Dashwood women. On the other hand, the presence of Jane and Elizabeth in the family circle seems to be necessary for its equilibrium, as Mr. Bennet remarks after his daughters' return from Netherfield: "the evening conversation, when they were all assembled, had lost much of its animation, and almost all its sense, by the absence of Jane and Elizabeth" (*P&P* 67). Another room that features several times in the novel is Mr. Bennet's library. It is his private space where he takes refuge from family life and he demands to be left alone there (*P&P* 125). Only very important matters are discussed there such as the marriage proposals (*P&P* 124, 385, 417). Elizabeth and Jane, being emotionally closer to each other than to the rest of the family, have their private conversations in the shrubbery (*P&P* 96) or in their room(s) (*P&P* 415).

Compared to Catherine Morland and the Miss Dashwoods, Elizabeth has not grown up in a psychologically supportive and emotionally stable household, nevertheless, she has managed to create a secure internal space for herself by means of relationships where her close family was deficient. For example, unable to confide in and to get sensible advice from her mother she turns to her aunt Gardiner who plays this role and who also provides her with the opportunity to observe a happily married couple. Moreover, at the time of the narrative she has already learnt, unlike her younger sisters and her parents, to behave in a socially proper way and has enough inner resources to make reasonable decisions for herself as well as for her mother, who has an uncontrolled emotional response to Lydia's elopement and her marriage. Elizabeth's relationship with her father is very affectionate, and, at the beginning of the novel, she seeks his recognition and enjoys their shared jokes, although "she had never been blind to the impropriety of her father's behaviour as a husband" (*P&P* 262). However, gradually, and with the help of Mr. Darcy's letter, she starts seeing his deficiencies as a parent and even tries to give him advice (*P&P* 256). In a way, even though she loves her father, she no longer looks up to him. Thus, by psychologically distancing herself from her father she gets ready to make a linear step into a socially higher and intellectually more challenging marriage.

In fact, Longbourn's lack of harmony, security, and order pushes Elizabeth to become emotionally self-sufficient and intellectually self-reflective. During the main action of the novel she has already out-grown her original home and her parents' family model. The story makes it evident that her role is not to preserve Longbourn for future generations, no matter how much her mother would like her to marry Mr. Collins. In this respect, and paradoxically, Elizabeth really is her mother's daughter as she is moving up the social ladder in a linear way and not starting a new cycle within her familiar sphere as Lady Catherine suggests she should wish to do (*P&P* 395).

Jane Bennet's wish to escape from her original home does not seem as radical as her sister's. The lack of harmony and balance in her family has rendered her tolerant and patient: she never prejudges people and expects the best of them (*P&P* 16). We also often see her ready to explain and justify people's acts, however strange, irrational, or mean they might look. For example, she is the only person to see a possibility of a misunderstanding in Mr. Wickham's tale of "his misfortunes" and to try to give some credit to Mr. Darcy's behaviour (*P&P* 95). This echoes the character and the situation of the "real gentleman," her uncle Mr. Gardiner, who is tolerant, helpful, and, like Jane, willing to see the situation in the best possible light when Lydia runs away (*P&P* 311), even though he grew up with two aggressive and eccentric sisters. In general, Jane plays the representative role in her family: she is the eldest and the prettiest daughter as well as the kindest and the most genteelly behaved; in other words, she is the best "product" of the Bennets. Thus, her successful marriage would represent the success of the whole family circle. Unlike Elizabeth, who acknowledges that her family makes her blush (*P&P* 111), Jane seems always loyal to it. Even in her most emotionally intense moment, after Mr. Bingley's proposal, she refers to her family saying: "to know that what I have to relate will give such pleasure to all my dear family, how shall I bear so much happiness!" (*P&P* 384). In time, however, she will learn to make her own choices for the best of her newly formed family and take some distance from her original home and her invasive mother by moving from Netherfield to Derbyshire close to her sister.

The original homes of the principal male characters of the novel are also important for the plot. The cases of Mr. Bingley and Mr. Darcy exemplify two opposite situations in relation to original homes. At the very beginning of the novel it is announced that Mr. Bingley is renting Netherfield Park (*P&P* 3). Later on we find out that his father "had intended to purchase an estate, but did not live to do it" (*P&P* 16), and that he left a large sum of money, "acquired by trade" (*P&P* 16) to his son; thus, at present, it is the role of the Bingley-son to found an inheritable family home. He is expected to climb the next step in his family's social ascension and a house—Netherfield Park or another—must be bought, not merely rented, in order to become a representative home-space. Therefore, Mr. Bingley's and his sisters' childhood home (which was undoubtedly a rented house) is non-existent—it is not relevant to the story: what is important to the narrative is Mr. Bingley's and, through him, his sisters' status of the "newly rich." The only perceptible attachment to his original space (*P&P* 3) would be Mr. Bingley's wish to buy a house in the North of England, though this may also be due to the influence of his model and best friend, Mr. Darcy from Derbyshire.

Mr. Darcy, on the other hand, comes from a strongly established and stable original home—Pemberley—although during the time of the novel's main action the household is in a transitory state: the old masters are dead and the young master is not yet married. We see that his childhood home becomes his adult home. In other words, he has to transform it into his adult home by founding a family and continuing the lineage. The house itself, in the image of its library, "has been the work of many generations" (*P&P* 41) which makes it a very cyclical space that has absorbed cyclical time. Every generation adds its contribution to the family picture gallery (*P&P* 277), where the visitor sees the succession of the young and the old and can feel the passing of time from the difference in clothes only. Mr. Darcy, by being brought up in this environment, has a moral duty to perpetuate this cyclicity, as Lady Catherine reminds him, when she thinks that the honour of the house is menaced by his inclination to marry Elizabeth Bennet with an uncle in trade (*P&P* 407). Concerning this point, Duckworth affirms that Mr. Darcy is "the responsible owner of a large house who is conscious of his responsibilities as trustee and who is aware (in Burkean terms) that he is not the 'entire master' but only the 'life-renter' of Pemberley" (*P&P* 129). Thus, by marrying Elizabeth, Mr. Darcy takes a courageous linear step and opens his ancestor's domain to inevitable, though not unwelcome, transformation.

This situation makes Pemberley only a half-home for Georgiana Darcy, for she cannot stay there on her own. Her situation is the opposite to Eleonor Tilney's, who had to stay in Northanger Abbey because her father kept his permanent residence there, as mentioned above. Georgiana's brother is her only family as well as her legal representative (together with Col. Fitzwilliam), and their relationship is presented as a very close one; however, she seems to spend a significant amount of time away from Pemberley with her "chaperon" lady and without her brother (Mr. Darcy, for example, spends two months at Netherfield Park while his sister is elsewhere). It is only after her brother's marriage to Elizabeth that Pemberley is said to become Georgiana's home (*P&P* 430). Thus, even though originally coming from a stable household, she is psychologically homeless up to the moment her brother settles down and provides her with a stable home again. Miss Bingley is in a similar "homeless" situation: she is obliged to follow her brother and adapt herself to his lifestyle and decisions and she has not got an inherited family home "waiting" for her. Moreover, as Miss Bingley does not seem to get married (*P&P* 430), she is left in this "homeless" and dependant state, as if punished for her greed, insincerity, and social ambition.

The first instance of temporary home in the novel reminds us of Marianne Dashwood's lengthy stay at Cleveland because of an illness, as it also marks a pivotal point in the narrative. Jane Bennet's visit to Netherfield Park lasts almost a week because of her common cold, though it happens in much more optimistic circumstances: she goes there to "hunt" for a marriage proposal and not to mourn a lost love, though she will have to pass through her phase of mourning later on. Moreover, she is still at a walking distance from her home, but not too close, and, thus, conveniently, her sister Elizabeth (and later, her mother with her other sisters) can visit and stay there too. Jane and Elizabeth's time at Netherfield is more difficult on a psychological level: the Bingley sisters, especially Caroline, show their hostility arising from jealousy and from their ambitious plans for their brother's marital future (*P&P* 40). The narrative mainly concentrates on Elizabeth and the development of her relationship to Mr. Darcy and how he realizes the degree of his interest for her (*P&P* 64), leaving sick Jane in the background. Numerous verbal jests that take place in the drawing room, like the one about weaknesses and faults of character (*P&P* 63), show their intellectual compatibility and betray their sexual attraction to each other that will be built on later in the narrative.

Another important temporary home space in the novel is Mr. Collins' house in Kent, where Elizabeth spends six weeks in Spring. Like Elinor and Marianne during the London season in *S&S*, Elizabeth goes through an important emotional upheaval in the course of her visit, but for her this disruption will turn out to be the beginning of a romance and not the end of it, as for Marianne. We see that from the very first days at Hunsford Parsonage Elizabeth is regularly looking for a private space, where she can be alone and think. Already during the first evening there, she is said "to meditate upon Charlotte's degree of contentment" in "the solitude of her chamber" (*P&P* 179). Though feeling uncomfortable in Mr. Collins' presence, she finds a way to inhabit this temporary space by isolating herself and plunging into her reflections. Noticeably, she needs more and more of this solitary time, especially when Mr. Darcy comes to visit his aunt bringing with him a considerable amount of "food for thought" with his enigmatic behaviour. He finds her alone at the Parsonage (*P&P* 199) when Charlotte and her sister have gone out; he "accidentally" meets her on her favourite solitary walks a few times (*P&P* 204) and during one of these walks, in the morning after his unsuccessful marriage proposal (*P&P* 217), he hands her his letter which threatens to render Elizabeth temporarily unsociable:

> After wandering along the lane for two hours, giving way to every variety of thought; re-considering events, determining probabilities, and reconciling herself as well as she could, to a change so sudden and so important, fatigue, and a recollection of her long absence, made her at length return home; and she entered the house with the wish of appearing cheerful as usual, and the resolution of repressing such reflections as must make her unfit for conversation. [...] She could think only of her letter. (*P&P* 231–232)

The passage shows how much psychological work Elizabeth has to do: she must re-arrange her world by re-assessing people and events, and, most importantly, her relationship with them. Thus, this temporary space, where Elizabeth is isolated from her sister in whom she could otherwise confide, absorbs the time of Elizabeth's internal transformation. Once she gets the occasion to talk to Jane two weeks later at Longbourn (*P&P* 251), she has already walked a good part of the path to a better understanding of the situation and has attained a degree of emotional distance as we can see from her joking comment about Mr. Darcy and Mr. Wickham that "one has got all the goodness, and the other all the appearance of it" (*P&P* 250). Later in the novel, there is another temporary space significant for Elizabeth and Mr. Darcy's relationship—the Inn at Lambton, which stands in a contrasting relationship to Pemberley as the Dream House.

In *P&P*, like in *MP* and *E*, the heroines' dream homes are central to the plots: Pemberley in *P&P*, Mansfield Park in *MP*, and Donwell Abbey in *E*. The particularity of these houses is that they are considered as very desirable homes by other characters and not only by the heroines, which corresponds to Bachelard's point of view as quoted earlier. Moreover, the images of these impressive houses mark the narrative from beginning to end; and the heroines, after some psychological difficulties that they have to overcome, are finally able to take possession of them.

Pemberley is the most striking dream home in Austen's fiction—it is depicted as absolutely perfect from the outside and the inside. If we do not count the gossip circulating during Meryton ball that Mr. Darcy has "a large estate in Derbyshire" (*P&P* 10), the first thorough information about the domain we get is from the conversation between Mr. Darcy, Mr. Bingley and Miss Bingley about family libraries in chapter 8 from which we see that the Bingleys qualify Pemberley as the most delightful place (*P&P* 41) and state that it is "more possible to get Pemberley by purchase than by imitation" (*P&P* 42), meaning that it is impossible to get it. Evidently, it represents a model home to the brother and the sister. Elizabeth, who is staying at Netherfield, and who has not shown any interest in her hosts' evening activities before this conversation, "was so much caught by what passed, as to leave her very little attention for her book" (*P&P* 42). She is intrigued by the image of perfection, possibly doubtful of its plausibility, but visibly, though yet unconsciously, attracted by what concerns Mr. Darcy. Noticeably, even though the accounts that she hears about Mr. Darcy are often negative in the first part of the novel, the image of Pemberley is never tarnished, not even by Mr. Wickham.

The following chapters of the narrative add some additional praise of Pemberley that elaborates its mental image in Elizabeth's mind (*P&P* 162, 267). However, the perception of it as a dream home comes during her visit there. We get a detailed description of Elizabeth's progress through the park and the view of the house itself, but most important is the account of her reaction to it:

> Elizabeth was delighted. She had never seen a place for which nature had done more, or where natural beauty had been so little counteracted by an awkward taste. They were all of them warm in their admiration; and at that moment she felt, that to be mistress of Pemberley might be something! (*P&P* 271)

The passage shows that Elizabeth forms her elevated image of the domain from what she sees. It is very different from the equivalent passage in *NA*. Catherine Morland has a vivid picture of her dream house/abbey preformed before she visits the actual place and has difficulties to see Northanger Abbey for what it really is. Elizabeth, on the other hand, gets impressed by the real, existing space, although the impact suggests that Pemberley corresponds to her idea of a perfect home. The fact, that Elizabeth is suddenly capable of picturing herself in this perfect space, seems to prove it.

During the visit she constructs her dream space from inside by absorbing room after room, the gardens and the park. She wishes to go further and "explore the windings" of the coppice-wood (*P&P* 280), as if she desired to discover every detail of the domain, what is hidden from a common outside eye, and so complete the image of her dream home. Noticeably, she does not see it only as a perfect house with grounds, but, by picturing herself as its mistress (*P&P* 272), she dreams it as a home space. Gradually, her emotional response to Pemberley shifts to the owner and, thus, she feels "a more gentle sensation" towards him (*P&P* 277), which is amplified later by Mr. Darcy's pleasant and respectful behaviour even though she is still embarrassed by their accidental meeting. During the visit, Elizabeth's mind is constantly travelling from the interior of the house to the outside view, as well as from the altering image of Mr. Darcy to the features of his domain. She is looking, hearing, thinking, and sensing. As M. McKeon remarks: "Austen narrates the entire experience consummately self-conscious of the interpenetrative relationship between vision and thought, object and subject, sensation and reflection, outside and inside" (711).

Elizabeth's dream home is pointedly compared with Mr. Collins' dream house—Rosings: "The rooms were lofty and handsome, and their furniture suitable to the fortune of the proprietor; but Elizabeth saw, with admiration of his taste, that it was neither gaudy nor uselessly fine; with less of splendour, and more real elegance, than the furniture of Rosings" (*P&P* 272). The passage conveys the feeling that Rosings, with its very expensive furniture and its "famous" chimney-piece worth eight hundred pounds (*P&P* 84) designed only to impress and to show off the owner's fortune, is considered as a "false" dream house. However, as Rosings corresponds to Mr. Collins' representation of perfection with all that is exaggerated and artificially "elegant" it continues to be a dream house for him.

Even though Elizabeth is apparently joking when she tells Jane that she fell in love with Mr. Darcy after having seen "his beautiful grounds at Pemberley" (*P&P* 414), the final chapter of the novel gives the impression that the acquisition of Pemberley is particularly important. Austen finishes her novel by enumerating and evaluating all the visits to Pemberley by their extended family members making it a central point in their life space (*P&P* 427–431). On the other hand, I can see an alternative explanation for the focus on Pemberley in the last chapter. Bearing in mind that Elizabeth's linear social leap brought some disequilibrium to the cyclicity of her family, the author sketches the new re-arranged functioning of the family and the re-organization of their connections in the last chapter. As I have discussed earlier, Elizabeth is psychologically ready to choose a different life trajectory and live up to it.

The first instance of a recently established adult home in the novel belongs to secondary characters: the reader has the possibility of taking a closer look at the adult home of a newly formed couple—Mr. Collins and Charlotte Lucas. Their situation is, however, a particular one, for even though they are living in the parsonage during the narrative (Charlotte moves in after their marriage at the beginning of the second volume), they are the legal heirs of Longbourn estate which will allow them to move up the social ladder and start their family cycle in that modest, but still trans-generational space. The announcement of Charlotte's pregnancy at the end of the novel (*P&P* 403) proves their union not to be barren, and symbolically "threatens" to settle their family in Longbourn for at least a couple of generations.

Bearing in mind that Mr. Collins has always known that he will inherit Longbourn, he might never have considered Hunsford as his permanent home. This might be the foundation of the particular way in which Mr. and Mrs. Collins relate to their Hunsford house. Mr. Collins lives at the front part of the house, making sure he can always have the view through the window that faces the main road as well as the garden. In fact, psychologically he is hardly inside the walls. Though proud of his "humble abode" and eloquent in its praise (*P&P* 177), his ambition pushes him out, towards Rosings, his dream house, because "of all the views which his garden, or which the county, or the kingdom could boast, none were to be compared with the prospect of Rosings, afforded by an opening in the trees that bordered the park nearly opposite the front of his house" (*P&P* 177–178). In a way, for him, Hunsford is more of a transitional space conveniently situated in the shade of his dream-house than an adult home.

Charlotte, on the other hand, lives in the very depth of the house, fitting things up there "with neatness and consistency" (*P&P* 178) and isolating herself as much as possible from her husband by pushing him out to work in the garden (*P&P* 177). Their sharing of space reflects the traditional gendered organization of space of the time: the male with the representative, public function occupies the outside space, while the female invests the inside space of the house and organizes the practical and decorative functioning of the household.[4] Charlotte's avowed goal is "only a comfortable home" (*P&P* 140) and she constructs one for herself in a rational and down-to-earth manner. Her psychological home is an adult home *par excellence*: practical, not based on "romantic" illusions and the very opposite of any other heroine's dream house.

Other adult homes of the novel are not depicted in a detailed way. Jane and Mr. Bingley purchase an estate one year after their marriage, thus, Mr. Bingley accomplishes his filial "duty" of establishing a permanent, inheritable family home. However, we only learn that the domain is situated 30 miles from its model, Pemberley, as I have already mentioned. The characters of Wickham and Lydia "suffered no revolution" (*P&P* 428) after their marriage. Being two dissipated personalities, they do not manage to create a stable home and, therefore, are said to be "always moving from place to place in quest of a cheap situation" (*P&P* 428–429). Thus, unlike Anne and Captain Wentworth in *P* who consciously choose the "sailor's" way of life without a permanent home, their relative homelessness is not freely chosen and might be considered as a logical outcome of their morally reprehensible behaviour and a reflection of their characters.

Mansfield Park

The question of the original home in *MP* is more complex than in other novels for the heroine has got two: her birth-home in Portsmouth and her adoptive home in Mansfield Park. The reader finds out nothing about Fanny's Portsmouth home at the beginning of the novel, except the size and the precarious economic situation of her family (*MP* 6). The narrator relates, however, her first reaction to her new home:

[4] Saggini and Soccio note, that "with the development of the proto-capitalist society, 'home' became ever more associated with the idea of an environment for women, especially of middle-class background, upon which female refinement and taste on the one hand and female virtue and modesty on the other could put their stamp" (3).

> The grandeur of the house astonished, but could not console her. The rooms were too large for her to move in with ease; whatever she touched she expected to injure, and she crept about in constant terror of something or other; often retreating towards her own chamber to cry. [...] A week passed in this way, and no suspicion of it conveyed by her quiet passive manner, when she was found one morning by her cousin Edmund, the youngest of the sons, sitting crying on the attic stairs (*MP* 16).

This passage conveys the idea that one needs to form an affective relationship with space in order to really inhabit it. Fanny has left her home and hasn't yet managed to find in Mansfield Park a substitute for it that could "console" her. The size and other aspects of the new space contrast too much with the space in which she spent the first ten years of her life. The words "crept" and "terror" indicate that she feels like a kind of intruder in this house. The grandeur of the rooms makes her feel exposed, thus she can't move "with ease" nor confidence. She is looking for places that are more nest-like, as Bachelard would put it, thus more isolated and secure: her room and attic stairs, where she can give way to her emotions. Edmund's attention to her, however, creates an affective relationship, and from this moment on "the place became less strange, and the people less formidable" (*MP* 19). It will however take much longer to form a strong bond with Sir Thomas and Lady Bertram, while with Julia, Maria, Tom, and Aunt Norris such a bond will never be created.

In *MP* Austen noticeably accords more attention to the house itself and the spaces around it than she does in her three previous novels. As Fanny grows up we see her gradually move her intimate space from the "little white attic" to the East room (*MP* 176–177), and she is said to have "worked herself into it" (*MP* 177). These are still her only places of real comfort in the house. Drawing on Bachelard's *The Poetics of Space*, J. Skinner comments:

> The East room or the attic as the places of Fanny's dreaming could of course be explained quite reductively: banished to an obscure corner of the house, the heroine has compensatory daydreams. And yet this aspect of the novel has a far greater importance: it reflects Fanny's gradual estrangement from Portsmouth and eventual identification with Mansfield. (11)[5]

[5] In his article "Exploring space: Constellations of Mansfield Park" Skinner analyses Fanny's instances of daydreaming which allows her to experience "intimate space at Mansfield" and thus "to strengthen her psychological ties with her new home" (8–9).

Thus, as she grows up, Fanny transfers her original home space from Portsmouth to Mansfield by physically and emotionally inhabiting the house, little by little, until it finally becomes hers at the end of the novel. Her physical, and non-voluntary return to her Portsmouth home will only confirm her total estrangement from that place. Moreover, as F. Johnston points out, her childhood space seems "too small for her" (199), in other words, she has outgrown her original home.[6] A. Duckworth goes even further, calling Fanny's Portsmouth home "a hostile new environment" and noticing the irony in its being "her previous home" (*MP* 72). I would suggest that Fanny feels as if her original space was hostile, because she is hoping to find her early childhood sensation of being at home with her family and, at the same time, dreading not to feel it: on her arrival in Portsmouth "Fanny was all agitation and flutter—all hope and apprehension" (*MP* 435). However, her psychologically lonely stay there proves that time and distance have led to a loss of this familial bond. This experience also means that Fanny drifts away from her original family and becomes a full member of the Bertram family as daughter (though she is a niece) and, after her marriage, as daughter-in-law too. Thus, by replacing her original home with Mansfield Park she simultaneously replaces her original parents with her aunt and uncle, and by doing so, she climbs up socially, as almost all the other heroines do.

Unlike other heroines (except Emma) who travel and stay in several different houses, Fanny only lives in her two homes—Portsmouth and Mansfield Park—during the main action of the narrative. The transformation of the status and importance of these spaces in Fanny's mind is traced throughout the story. At different times, these two houses have different roles. Portsmouth passes from the original home to a temporary home, and then, to no home at all, whereas Mansfield Park not only replaces her childhood home, but also carries features of a temporary home, the dream house and, finally, an adult home, as I will show below.

Mansfield Park in *MP* is the most polyvalent home space in Austen's novels. The house itself corresponds at different moments to all four categories: it is an original home for the Bertram children and, later, Fanny; it is a transitory home for Fanny and Edmund while they are still single; it is a dream house for Mary Crawford and Fanny when she is away; and, finally,

[6] In her article "Public and private space in Jane Austen" Johnston juxtaposes Fanny's first impression after her arrival at Mansfield, when she was "disarmed by the size of Mansfield" (199) and her adult impression of the smallness of her Portsmouth home.

it becomes an adult home for married Fanny and Edmund. As an original home it has a different character for Maria, Julia, and Tom Bertram than it does for Edmund and Fanny. For the Miss Bertrams and Tom it represents an oppressive household marked by the unwelcome authority of their father and the psychological incompetence of their mother. They try to disturb and shake up its internal order by the theatricals, and after their project fails they escape it as soon as they can: Maria rushes into an unhappy interested marriage taking Julia with her, while Tom seems to stay away as often as he can. Edmund's relationship with his childhood home is less tense. Being as conservative and principled as his father, though less authoritarian and more affectionate, he is the proper heir who is able to perpetuate the cyclicity of an old, strong, and stable English landowning family.

There is one more significant temporary home space in the novel (next to the temporary aspects of Mansfield Park and Portsmouth) which concerns the house that is geographically as well as socially close to the original home of the heroine—the parsonage, where Mary and Henry Crawford stay for several months with their sister. Arguably, any parsonage is a temporary space by definition, for it cannot really be inherited but is bestowed on a new clergyman once the old one dies or moves; and after Henry Crawford's elopement with Maria, the Grants presence is no longer morally accepted there (*MP* 542), therefore they find a convenient arrangement with another parish that justifies their living in London. In addition, the Mansfield parsonage is an alternative space to Mansfield Park: it represents a freer and more spontaneous household, which, during their visit, is run according to the habits and whims of the Crawford siblings. Symbolically, both houses "though scarcely half a mile apart, were not within sight of each other" (*MP* 78) thus marking a gap in their social and moral character. Both spaces are, however, closely connected, because the young characters are interested in each other and regularly walk to each other's home. Mansfield Park absorbs some of the spirit of the parsonage with the theatricals and the ball, while Mary Crawford acknowledges to Fanny that, though unusually quiet, the country season at Mansfield was the happiest time for her (*MP* 245). The parsonage for the Crawfords is also a very cyclical space: while staying there, they explore different possibilities of re-directing their lives by "trying out" romantic relationships with all the brothers and sisters of Mansfield Park, including Fanny. When their respective plans with Fanny and Edmund fail, and their choices—Henry's elopement with Maria and Mary's first plan to get Tom Bertram, the heir—turn out to be unsatisfactory, they move back to London, the polyvalent city space, which is more suitable to their unstable characters.

Mansfield Park, as I have already mentioned, is a multifunctional home space. The domain is perceived as a dream home first by Mary Crawford, and later, during her Portsmouth visit, by Fanny, though what is comprised in this notion differs for each of them. Both of them are taken by surprise in realizing how much they appreciate the place, but while Mary will have to renounce the dream home, Fanny will gain it.

Mary Crawford becomes interested in Mansfield Park and its heir Tom immediately after her arrival at the Parsonage for she thinks marrying Tom would consolidate her social status. The narrator ironically relates her "feelings":

> Miss Crawford soon felt, that he [Tom Bertram] and his situation might do. She looked about her with due consideration, and found almost everything in his favour, a park, a real park five miles round, a spacious modern-built house, so well placed and well screened as to deserve to be in any collection of engravings of gentlemen's seats in the kingdom, and wanting only to be completely new furnished—[…] an agreeable man himself—with the advantage of being […] Sir Thomas hereafter. (*MP* 55)

This passage, which resembles a real estate advertisement for a perfect house, indicates that she perceives Mansfield as a dream home in a social sense. Mary is not touched emotionally, but she considers it to be a highly desirable domain according to social norms: what should be desirable for an ambitious lady that seeks to make an advantageous marriage and gain a higher social status. Tom matters only as the heir of the house and of the title. Thus, Mary's image of a dream home is artificially, and not emotionally constructed, as it is with Austen's main heroines. Moreover, later, when she acknowledges to Fanny her surprise at feeling happy in a country house (*MP* 245), she points out that she "can even suppose it pleasant to spend *half* the year in the country, under certain circumstances"—in other words, if she can have a lot of social engagements and balls (*MP* 245). Thus, her real dream home comprehends not only a perfect house, but also a splendid lifestyle in a varied, amusing, and rich social circle.

However, as it turns out, Mary Crawford is yet another case of female "homelessness" in the novels. She experiences a similar situation to that of Caroline Bingley and Georgiana Darcy—being dependant on her brother and his decisions. Mary has a choice between living with her married sister, Mrs. Grant, or her single brother, but she must adapt herself to their lifestyle, even though it does not suit her, as the following passage illustrates:

> Mrs. Grant's […] chief anxiety was lest Mansfield should not satisfy the habits of a young woman who had been mostly used to London.
>
> Miss Crawford was not entirely free from similar apprehensions, though they arose principally from doubts of her sister's style of living and tone of society; and it was not till after she had tried in vain to persuade her brother to settle with her at his own country-house, that she could resolve to hazard herself among her other relations (*MP* 47).

The reader will find out at the end of the novel that Mary remains living with her widow sister in London after the death of Mr. Grant (*MP* 543), and like the equally greedy and manipulative Miss Bingley, possibly never marries.

Fanny's attachment to Mansfield is of a very different kind: she gradually gets accustomed to this place, first accepting it as her home space, and then, realizing that it is the house in which she would like to remain. She starts dreaming of Mansfield Park during her stay at her former home in Portsmouth: "The elegance, propriety, regularity, harmony—and perhaps, above all, peace and tranquillity of Mansfield, were brought to her remembrance every hour of the day, by the prevalence of every thing opposite to them *here*" (*MP* 453). Even though the narrator describes Fanny as remembering, the reader knows that Mansfield has not always been peaceful, harmonious, and without any breach of propriety. Therefore, we can deduce that Fanny is (re)creating in her mind a picture of the ideal Mansfield Park—her dream home. And when she returns to Mansfield, she finally feels there in her rightful place—useful, loved, and, thus, happy (*MP* 533).

Noticeably, Austen presents the adult home of Fanny and Edmund more as a psychological home space "of affection and comfort" (*MP* 547) than a specific house. This conveys the impression that their home is first of all their union and companionship, as much in Thornton Lacey as in Mansfield Parsonage. The last line of the novel seems to indicate that their real home space comprehends not only the Parsonage, but also Mansfield Park. Thus, the great house undergoes one more transformation: after having been a childhood home, temporary space, and dream house, it becomes Fanny and Edmund's adult home.

Emma

In *Emma* the childhood home—Hartfield—is at the centre of the novel. The text refers to it as the Woodhouse family seat and the guarantee of Emma's social, financial, and intellectual independence. However, at

21 years old Emma is left alone to run the household. From the very first pages of the novel, the big family house is getting emptier and becoming sadly too "big" for just Emma and her ageing father (*E* 5). In her childhood Emma lost her mother, who was, according to Mr. Knightley, "the only person able to cope with her" (*E* 38); and he adds when speaking to Mrs. Weston that "ever since she was twelve, Emma has been mistress of the house and of you all [her father, sister, and governess]" (*E* 37). Thus, Emma's childhood home has very early been transformed into an adult home, especially given that, at least during the first half of the novel, she does not even consider marrying and abandoning her house together with her status of mistress.

Unlike other heroines, Emma does not have to leave her home, and she is the only one that has the power of decision concerning household affairs. The narrator suggests that this situation is not the most advantageous for a young woman in a psychological sense: "the real evils indeed of Emma's situation were the power of having rather too much of her own way, and a disposition to think a little too well of herself" (*E* 3). Consequently, Emma does not have to resist her family's ways or plans, as Elizabeth Bennet or Fanny Price do, since it is rather she who is presiding, suggesting, and imposing. This marks the vulnerability of Emma's original home: for her home is first of all herself. She also takes it upon herself to continue the ancestral cycle and she is very proud of her family history, as the reader can see from her soliloquy about Mr. Elton and his marriage proposal (*E* 147).

On a phenomenological level, her home resembles a shell. Living in a shell means first of all living alone (Bachelard, 123), but even though she does not live there on her own physically, in a psychological way she is alone in her original home "suffering from intellectual solitude" (*E* 5). Her house recalls a shell also because it is perfectly comfortable and adapted to her. She is moreover, morally attached to it by her old hypochondriac father and cannot leave it. The solution to this impasse will be to break this shell by enlarging the household, as we will see later.

On the other hand, Emma seeks to fill up her emptied home space after the marriage of Miss Taylor. Her new friend Harriet, though not intellectually nor socially her equal, becomes almost an inmate at Hartfield, while Mr. Knightley is a daily visitor, and Emma is also very happy to have her sister's family there for holidays (*E* 5). The final accomplishment of this project of filling up her home space will be Mr. Knightley's moving in after their marriage: an extra-ordinary event for the time, when normally women left their homes to live with their husbands. Thus, like Mansfield Park in *MP*, the Hartfield house changes its status during the novel.

In *E*, as in *P&P*, we also have an example of a stable trans-generational landowning family home—Donwell Abbey. However, its owner, Mr. Knightley, who grew up in this house, has already turned it into his adult home. Unlike Mr. Darcy, he is less under pressure to get married and produce an heir, for he has got a brother with several children, and so the continuity of the Knightley family is assured. On the other hand, as in Mr. Darcy's case, the stability of the original home reflects Mr. Knightley's psychological stability, logical reasoning, honesty, and emotional control.

As a contrast to Mr. Knightley Austen presents Frank Churchill, a half-orphan, who had to leave his childhood home when he was about two years old and who was adopted by his aunt. While staying in contact with his biological father Mr. Weston, Frank is an heir of the Churchill's fortune. The duality of his original home is reflected in the duplicity of his behaviour: he gets secretly engaged to Jane Fairfax and tries to hide it from Highbury society by flirting with Emma. Moreover, he is shown to be a superficial and quite selfish person guided by his emotions. On the other hand, Frank seems to be drawn to his early childhood space: he wishes to see the former Weston family home (*E* 212), and he chooses a bride who is native to Highbury and who, like him, had been sent away from there in her childhood. In a way, by doing so, he unconsciously seeks to re-enter his original family circle while, at the same time, staying a decent adoptive son to his aunt and uncle. Thus, the dualism of Frank's character can be understood as originating from the dualism of his childhood home experience.

The houses that are entailed, rented, or are to be abandoned for financial reasons have a temporary character and the relationships of the characters with these home spaces are therefore emotionally intense. This is the case with Norland, Longbourn, Netherfield, and Kellynch Hall, as discussed in the first part of this chapter. However, characters can also create a relationship that is strongly marked on the psychological level with homes that are not their original/childhood ones. The perfect example is Mrs. Bates' home for Jane Fairfax in *E*.

Jane Fairfax, like Fanny Price, has got two childhood homes. However, in her case, both homes are clearly temporary, for she is an orphan brought up by her remaining family and her guardians, the Campbells, to become a governess (*E* 175). Jane has never really lived with her grandmother, since after her mother's death she "had belonged to Colonel Campbell's family, and had lived with them entirely, only visiting her grandmother from time to time" (*E* 175). Thus, this is the first time that she comes back to Highbury for such a lengthy stay, knowing that she won't be returning to live with her foster family any more. Jane apparently views her stay with the Bateses as a transitional time to improve her health (and as her mother died of consumption [*E* 174], Miss Bates shows a lot of concern for her niece's health) as well as to give herself a few additional months of leisure before entering the "job market." However, secretly, she is biding her time and waiting for Frank Churchill's situation to change. Jane is repeatedly shown as too elegant, too delicate and too talented for her grandmother's modest home and income. She can hardly fit in Highbury and in Mrs. Bates' small flat as, symbolically and physically, the pianoforte hardly fits in her aunt's drawing room. Jane seems to have done everything in her power to be eligible for the lifestyle of a lady by achieving proficiency in music and elegance. The narrator convinces the reader that she deserves to continue living in the "best" social circles with her numerous personal qualities and accomplishments and "prepares" the reader for the positive outcome of Jane's situation. However, the only thing Jane can do at this point of the narrative, is wait in the psychological security of her family's tiny home space for some lucky event to transform her destiny. She is loved there, but she does not feel at home. The close intimacy with her aunt and grandmother renders the concealment of her relationship with Frank difficult and especially oppressing. When in doubt about Frank's love, she seems to find her small temporary home as little comfort, possibly even suffocating, and, like Marianne Dashwood in *S&S*, she is "seen wandering about the meadows, at some distance from Highbury" (*E* 426).

As I have already mentioned, in *E* Austen presents us with one of the most significant dream homes of her novels. Interestingly, Donwell Abbey acquires its features of a dream home gradually as the narrative advances, although it never forms such a striking picture as Pemberley or Northanger Abbey. In a way, it is always in the background of the action and comes to the foreground only in a couple of passages. In the first volume of the novel, while Emma is preoccupied with getting Harriet married to Mr. Elton, Donwell is mentioned only as home to Mr. Knightley without any further implications, although we get a sense of how important and highly esteemed Mr. Knightley is and that he is a large landowner with tenants (*E* 62, 147). Arguably, the first time that the reader's attention is drawn to the significance of Donwell Abbey is during the conversation between Mrs. Weston and Emma during the dinner party at the Coles, when Mrs. Weston suspects Mr. Knightley to be in love with Jane Fairfax (*E* 242–243). Besides the first outburst of Emma's unacknowledged feelings for Mr. Knightley, the dialogue also marks Donwell Abbey out as a desirable domain not worthy, in Emma's opinion, of less socially elevated people: "Jane Fairfax mistress of the Abbey!—Oh! No, no;—every feeling revolts,"—she exclaims (*E* 243). However, we get a proper tour of the domain only in Summer, during the strawberry party, when all the guests get a chance to admire it (*E* 390–391). Symbolically, it is viewed on the Summer solstice—at the "highest" point of the year—surrounded by ripe fruit, which evokes fertility, self-sufficiency, and dream-house quality. Finally, the question of Donwell arises when Mr. Knightley and Emma get engaged and he decides to move to Hartfield while Mr. Woodhouse is alive. Officially Emma becomes its mistress, though not immediately residing there; and we know that eventually she will occupy the envied position of Mrs. Knightley, who is allowed "to invite what guests she pleases to Donwell" (*E* 385). Thus, by gaining Donwell and by staying at Hartfield at the same time Emma remains within her familiar space, but, bearing in mind her character and her wish to be the central point of her world, having the two most prominent houses of the neighbourhood would correspond to her notion of dream home.

As with the Collinses in *P&P*, the reader can also observe a recently married couple in an independent adult home in *E*. Randalls is a newly acquired estate that enables Mr. Weston to marry and, eventually, start a new family after many years of work and widowhood. He chooses Miss Taylor, an ex-governess, a mature, steady, experienced woman. Thus, it is not surprising that their household seems to form itself naturally, harmoniously, without any interior struggle. The only "problems" that they meet concern Frank Churchill and his disconcerting filial behaviour. In short, the acquisition of Randalls for Mr. Weston and, indirectly, for his wife, is a natural consequence and reward for their hard-working years. It is also a lucky linear step: like Mr. Bingley in *P&P*, though at another level, they thus accomplish their social ascension.

On the other hand, *E* provides one more "homeless" female character, though in different circumstances than Georgiana Darcy and Caroline Bingley in *P&P* and Mary Crawford in *MP*—Harriet Smith, who is a natural daughter of an anonymous person and, thus, does not have even a deficient original home; who grew up in Mrs. Goddard's boarding-school (*E* 22), continues to board there even though she has completed her education, and risks staying there for good if she does not find a man to marry. Ironically, Emma almost succeeds in stopping her from marrying a man who can give her a secure home.

In conclusion, the situation of the principal houses in *E* is quite different from the other novels for neither the heroine nor her husband-to-be need to go away from their original homes. Mr. Knightley, as I have mentioned, is mature enough to invest Donwell as his independent adult home space and his moving to Hartfield after the marriage is only temporary. Similarly, Emma, first considers herself as the unique, thus unmarried, future mistress of Hartfield (*E* 90–91). Therefore, their marriage only officializes, strengthens, and confirms the status of their already adult homes, especially as even Emma's and Mr. Knightley's living together is partly tested by his being a very frequent visitor at Hartfield (*E* 8), often fully acquainted with all its events and always giving his opinion on Emma's acts and decisions, as we see throughout the novel.

Persuasion

From the very beginning of *P* the importance of the family cycle for the Elliots is emphasized, for it is said that Sir Walter Elliot's favourite occupation is to read and re-read his own family history in the Baronetage (*P* 3). Moreover, from these very first pages Anne Elliot stands out from "all the Marys and Elizabeths they [the ancestors] had married" (*P* 4), as if she is being pushed out of that cycle by her Christian name. Sir Walter Elliot is said to depend on his eldest daughter Elizabeth to perpetuate the lineage and marry well and considers his other two daughters of "very inferior value" (*P* 5). However, as in *P&P* the main tension arises from the fear of losing their family home—Kellynch Hall—which is entailed on a distant relative (*P* 4), and the family is deep in debt (*P* 10).

The relationship of Anne Elliot with her original home is ambiguous. She must have had a happy childhood in Kellynch Hall, for it is said that she had a very caring mother (*P* 4–5) whom she resembled, according to Lady Russell (*P* 6); however, from the age of fourteen, after her mother's death, her status in the family has changed: "her word had no weight; her convenience was always to give way;—she was only Anne" (*P* 6). Nevertheless, like the Dashwood women, she stays attached to the house itself and the area, even though her life there with her family is no longer satisfactory. After her unsuccessful plans of retrenchment that were not accepted by her father and sister, she hopes at least to "still have the pleasure of sometimes seeing the lawns and groves of Kellynch" (*P* 15). This shows that Anne is not ready to leave her childhood home and she feels the moving out of it as a banishment. Her original home still carries the mark of her happy, though far-away, past which makes her cling to it. She chooses to stay in the neighbourhood in order to mourn her loss during the time when her home becomes somebody else's home, instead of filling her life with new experiences as her father and sister do. Moreover, the mourning of her childhood home turns into a more general mourning and reassessing of her past, when Captain Wentworth arrives to stay at the rented Kellynch Hall and, thus, occupies the space that used to belong to her.

In contrast to the failing Kellynch Hall, The Great House at Uppercross is presented as a comfortable and hospitable family space and the Musgroves, like the Morlands in *NA*, as a united, friendly, though not cultivated family (*P* 43). The warmness of their relationship with each other stands in opposition to the coldness of the Elliot family relationships and Anne envies them "that seemingly perfect good understanding and agreement together, that good-humoured mutual affection, of which she had known so little" (*P* 43–44). The original home of another important character —Captain Wentworth—is not mentioned in the novel, but he is said to "come to Kellynch as to a home, to stay as long as he likes, being as thoroughly the object of the Admiral's fraternal kindness as of his wife" (*P* 79). This makes clear that his ties with his sister were close and affectionate, thus forming another contrast to Anne's situation in her family. Like Edward Ferrars, Captain Wentworth finds a home in his kinship relations: first he stays with his sister, and later, after the accident in Lyme, he goes to his brother. He will establish his own adult home only after marriage, as I will discuss later.

As in *S&S* and *P&P*, a temporary home *in extremis* marks one of the pivotal points in the plot. Louisa Musgrove, like Marianne Dashwood, spends some time between life and death in Harville's house in Lyme, and, again similarly to Marianne, gets out of this trying state psychologically transformed. During her slow recovery, the small and crowded house of the Harvilles becomes a temporary home to her, and it also initiates the pivotal point in her life—her marriage to Captain Benwick. Thus, it is the space/time of Louisa's physical and psychological transformation. Moreover, this episode is crucial to the main plot line: by getting engaged to Benwick, she liberates Captain Wentworth for Anne.

Anne Elliot finds a temporary home with her hypochondriac sister at Uppercross Cottage, after her original home is rented and before she is to be "banished" to Bath. It is a familiar space where she used to stay regularly and "she knew the ways of Uppercross as well as those of Kellynch" (*P* 39). In a way, she finds a substitute home to soften the transition from her old life to the new one in Bath. The lodgings that the

Elliots rent in Bath also represent a temporary space, at least what is hoped to be a temporary space for Anne and Elizabeth, who should quit it once they marry. And though "Anne entered it [the house in Bath] with a sinking heart, anticipating an imprisonment of many months" (*P* 148) we do not get the feeling that she is permanently fixed there. From the very first day of her arrival, the reader is kept in suspense between her two potential fiancés—Mr. Elliot and Captain Wentworth—who would be able to provide her with a permanent adult home. Noticeably, in both cases she would gain a closer contact with her original home. If she married Mr. Elliot she would become the heiress of Kellynch Hall after her father's death; whereas, if she married Captain Wentworth she would be staying there as a visitor as long as the Crofts rent it. During the last part of the narrative the reader follows Anne's emotional path towards her final conviction of Captain Wentworth's being the only man she loves and her anxious expectation of a response to her feelings. Thus, Bath is also a linear space where the most important decision concerning the future life is made and where the evolution of Anne's and Wentworth's emotional history is finally completed.

Like Marianne, Anne Elliot in *P* does not possess her dream home. In general, this notion applies to Kellynch Hall, which is, at the same time, Anne's original home. However, the house is not only the lost original home, but also a "prize" to win in case of marriage with the presumed heir—Mr. Elliot. Thus, Elizabeth Elliot indirectly competes for it with Anne, till Anne clearly shows her disinclination to marry Mr. Elliot, and he turns his attention elsewhere. The narrative is divided into two parts in this respect: the first part features Kellynch Hall as lost original home space, while in the second part, after the re-appearance of Mr. Elliot, it becomes the most desirable house to acquire, at least in the eyes of Elizabeth Elliot and Lady Russell. Even to Anne it is a tempting prospect as we see from her reaction to Lady Russell's depiction of her possible future as "Lady Elliot":

> For a few moments her imagination and her heart were bewitched. The idea of becoming what her mother had been; of having the precious name of "Lady Elliot" first revived in herself; of being restored to Kellynch, calling it her home again, her home for ever, was a charm which she could not immediately resist (*P* 174).

The extract shows that getting back the house is equal to becoming what her mother once was, in other words it means getting her lost mother back by occupying her place in a symbolic way. For a moment, Anne is carried away by these dreams of the possibility to re-find her happy childhood household, to re-live the lost past, and to extend it into the future by transforming Kellynch into her adult home. However, she rapidly overcomes this temptation, and renounces this symbolic image. Anne is sufficiently psychologically mature, to be able to move away from her family's cyclical trajectory that she was persuaded to follow eight years ago. Becoming Captain Wentworth's wife means "no landed estate, no headship of a family" (*P* 272), but to Anne these have lost their importance and "she gloried in being a sailor's wife" (*P* 275) and being without a permanent let alone dream home.

The reader witnesses a very different case at Uppercross Cottage. Certainly, Charles Musgrove and his family are living there only temporarily, before he inherits his parents' estate. However, it is his first adult home and the childhood home for his children, which makes it a psychologically significant space. The troublesome and hypochondriac personality of his wife Mary, transforms their home into an imbalanced and stressful place. Therefore, they need an external balancing force—Anne—to make their household hold together at critical moments, such as their son's accident or Mary's fits of imaginary illnesses. Noticeably, the character of Anne Elliot, recalls the calmness and good common sense of Mrs. Weston in *E*; and even though we are not given any precise indications in the narrative concerning Anne Wentworth's adult home, we can infer that these personality traits will be a solid psychological foundation for a well-balanced adult household.[7]

The Watsons and *Sanditon*

Both Austen's unfinished novels, *The Watsons* and *Sanditon*, provide interesting examples of childhood homes too. *W* starts with the main heroine Emma Watson's return to her family home from "the care of an aunt who has brought her up" (*W* 79). Emma's situation seems to be similar to Jane Fairfax's in *E*: she is poor, motherless, and has spent her childhood away

[7] Billi sees in Austen's "praise of the glorious English Navy" (70) and the hint of a possible war that the house of Anne and Frederic Wentworth will be "a ship, freely ploughing the waves at sea" (71). After 1816 (the completion of *P*) the British Navy participated in various small conflicts and, thus, it is possible that Anne would have spent her time with her husband on a ship. This was the case for Mrs. Croft who lived on a ship with her husband, and in spite of being "confined" she claims that "the happiest part of [her] life has been spent on board a ship" (76).

from home. However, she has got a home with her father and sisters in it, even though obviously deficient: her father, like Mr. Woodhouse, is said to be sickly and dependent (*W* 79), and her sisters are competing for the same man (*W* 81), which might remind the reader of Maria and Julia in *MP*. Therefore, judging from the first pages of this unfinished text, Emma Watson does not have a stable and reassuring original home experience, and all her hopes would presumably be directed to finding a suitable husband and thus escaping the deficient family home.

Differently, in *S* the reader is immediately given a positive picture of a harmonious and reassuring home space. The Heywoods are described as attentive, competent, and kind. Their household with 14 children gives an image of self-sufficiency: the parents never leave home and have everything they need in their domain (*S* 149). Thus, it is a stable and cyclical home space, which constitutes the background of the presumed heroine Charlotte Heywood. Her initial situation resembles the situation of Catherine Morland in *NA*, although she seems to be a more down–to-earth character. However, even though happy at home, Charlotte is gently pushed by her parents (*S* 145–150) to go away and experience the outside world. Thus, Mr. Parker's accident turns out to be very handy.

Mr. Parker, on the other hand, has broken the cyclicity of "the house of his forefathers" (*S* 155) by building a modern house for his family. The original home is "well fenced and planted" and sheltered in "a dip within two miles of the sea" (*S* 155): we get a sense of roundness, security, stability, and self-sufficiency. This reminds Charlotte of her own home (*S* 155) and represents traditional values associated with a trans-generational family house. Mr. Parker makes a linear leap by getting out of the ancestors' house "built in a hole" (*S* 156). His attitude towards his original home space is very different from other characters'. It is the first time in Austen's work that a wealthy gentleman chooses to abandon the old family home for a modern newly-built house in a very unusual, un-sheltered place, and without expressing any wish to re-integrate the ancestral cycle later.[8]

[8] Duckworth suggests that Mr. Parker's actions are doomed to fail, as prefigured by his carriage crash in the beginning of the novel. He notes:

> Mr. Parker has left behind a life (as well as a garden) that was socially formal, yet in tune with the natural rhythms of the country. More important, he has left behind a society that provided protection and support from contingencies, whether natural, like the storm, or social. In flirting with his commercial ventures, he may be inviting "real danger". It seems probable that had *Sanditon* been completed the bubble of Mr. Parker's "speculation" would burst. (215)

In fact, Mr. Parker really creates his new home, gives it an up-to-date glorious name, and tries to make it very different from the ancestral home of his family: "Trafalgar House, on the most elevated spot on the down, was a light elegant building, standing in a small lawn with a very young plantation round it, about an hundred yards from the brow of a steep, but not very lofty cliff" (*S* 160). The passage gives an impression of an unprotected house that is dangerously exposed to natural forces. Moreover, even though needing to buy his vegetables from the old house and being obliged to equip his children with large bonnets and parasols when they play in the grounds, Mr. Parker rejoices at being able to admire "the grandeur of the storm" (*S* 157). By constructing his new family house in this anti-traditional way, Mr. Parker is trying to make a point of being a romantic pioneer. Despite the fact that we do not know how his new life in this modern house will evolve, we can feel Austen's irony when she compares the traditional space with this ultra-modern and romantic one. In addition, noticing the superficiality of Mr. Parker's reflections and ironic narrator's comments on his new house where there is no shade nor shelter from the storms (*S* 156–157) the reader gets the feeling that the author is not on her very modern character's side.

Connecting Spaces and Thresholds

After being "suddenly roused" by the doorbell, Elizabeth, in amazement, sees Mr. Darcy enter the room (*P&P* 211). The entrance is followed by his marriage proposal, which turns Elizabeth's surprise into an "astonishment beyond expression" (*P&P* 211). This visit marks the main threshold in their life stories: from this critical point on they will begin their gradual emotional progression towards each other. The passage perfectly illustrates Bakhtin's point that the threshold is "highly charged with emotion and value [...] whose fundamental instance is the chronotope of *crisis* and *break* in a life" (248).

In the novels, these thresholds are represented by images of windows and doors, which, beside their mimetic role, carry out different functions, such as connecting, disclosing, limiting, separating, and which move the narrative towards and over critical points. As J. Pidduck comments when writing about Austen film adaptations: "in a cycle of texts organized

He adds, that the word "speculation" appears several times in this part of the text and claims that it must have a pejorative meaning (216).

around the viewpoint of bourgeois female characters who are both actually and metaphorically 'housebound', windows and doors where arrivals and departures occur provide focal points of narrative interest" (382). The doors and windows in films thus highlight the perspective of the female, who is mostly confined to a passive life indoors.

As in the films the novels give us images of women, sewing, embroidering, writing letters, or just sitting in their drawing rooms until a visitor comes and shifts a stagnant situation. That is what happens to Fanny Dashwood in *S&S* when she is "sitting all alone at her carpet-work, little suspecting what was to come" (*S&S* 293) before Nancy Steele comes in and tells her the shocking news about Lucy's and Edward's engagement; and to Jane Bennet, in *P&P*, who "resolutely" keeps "her place at the table" (*P&P* 369) when Mr. Bingley returns to Netherfield almost a year after their first meeting and visits Longbourn. Likewise, though less dramatically, Anne Elliot and her sister Mary are "sitting quietly at work" when the Miss Musgroves come and take them for the walk to Winthrop during which Anne starts feeling jealous of Louisa and Captain Wentworth (*P* 89). Being limited by the rules of propriety, women cannot always go out and seek for society themselves.[9] This situation is especially striking in *S&S*, after the Dashwoods move to Barton cottage and have to lead a much simpler life-style: "their visitors, except those from Barton Park, were not many; [...] and she [Mrs. Dashwood] was resolute in declining to visit any family beyond the distance of a walk" (*S&S* 48). On the other hand, some women are capable of finding ways to expose themselves socially as much as possible. In *NA*, for example, we find a comical passage about Isabella Thorpe's choice of her sitting place in the Pump-room: "'This is my favourite place,' said she [Isabella], as they sat down on a bench between the doors, which commanded a tolerable view of every body entering at either, 'it is so out of the way'" (*NA* 145). A place between the doors is a very exposed place, thus, it is the appropriate place for such an opportunist as Isabella Thorpe.

These thresholds, represented by doors and windows, can, however, be significant too, though in a different way, to male characters, as the example of Mr. Darcy's marriage proposal above illustrates. Therefore, in what follows I take into account both female and male characters when analysing the relationships to significant thresholds.

[9] See Le Faye, 113–114.

Doors

Bachelard sees "doors" as a very dream-productive phenomenon. In his chapter "The dialectics of outside and inside" he comments:

> The door is an entire cosmos of the Half-open. In fact, it is one of its primal images, the very origin of a daydream that accumulates desires and temptations: the temptation to open up the ultimate depths of being, and the desire to conquer all reticent beings. The door schematizes two strong possibilities, which sharply classify two types of daydream. At times, it is closed, bolted, padlocked. At others, it is open, that is to say, wide open (222).

In the novels, however, doors seem to be kept closed most of the time: the entrance doors, the doors to rooms and the garden gates limit and divide space into distinct cells that compose the house and/or the entire domain. There is of course a practical reason for keeping doors shut as this preserves the warmth of the fire in a chimney. References to fire are, moreover, quite frequent in the novels. Noticeably, a fire is often described as too warm or not warm enough, too close or too sheltered, or even as absent as in Fanny's room in *MP.* C. Johnston, discussing public and private spaces in Austen's novels, suggests that windows and fires are two important reference points in some rooms: "the heat from a fire" and "the view from a window" (203). For example, she points out that the balance between fire and window is disrupted in *MP*: 11 "windows," correspond to 27 "fires" making "the drawing room too hot for a sensitive newcomer such as Fanny Price" (204). I would extend this antagonism to the phenomenon of doors as well as windows, for both are opposed to fire in the novels, as I will show later.

Furthermore, the act of opening the door or of having it opened by somebody is charged with significance, and these acts often carry an element of surprise, even when they have been expected by those in the room. One of the most astonishing entrances is that of Mr. Darcy in *P&P* when he visits Elizabeth alone for the first time in Hunsford Parsonage: "the door opened, and to her very great surprise, Mr. Darcy, and Mr. Darcy only, entered the room" (*P&P* 199). His aunt's visit to Longbourn is hardly less surprising for the Bennets and rather more aggressive: "the door was thrown open and their visitor [Lady Catherine] entered" (*P&P* 389). It seems from this passage that Lady Catherine does not wait for the servant to open the door, as, normally, the passive form would suggest, but does it herself, thus demonstrating her anger and her direct, authoritarian character.

There are similar instances of surprising entrances into closed rooms in other novels. Anne Elliot in *P* gets settled "by one of the fires in the octagon room" waiting for their cousin Lady Dalrymple, when the door unexpectedly opens and Captain Wentworth walks in (*P* 197). Here we see a clear opposition between a fireplace, which represents the warmth, the cosiness and thus, comfort and predictability, and the opened door—that stands for the outside, the unknown, the unexpected. Captain Wentworth's entrance will turn out to be very important to Anne because his hectic behaviour during the concert will allow her to understand his real feelings (*P* 207). Later in the narrative, Wentworth will open the door unexpectedly again, to give Anne his letter containing his marriage proposal (*P* 257) which will put an end to Anne's eight years of mourning.

Emma, in turn, gets surprised when opening the door into her own drawing room: "she opened the parlour door, and saw two gentlemen sitting with her father—Mr. Weston and his son" (*E* 204). At this point of the novel, the incident illustrates another of her false presuppositions, for she had calculated that Frank would come later. On the other hand, it is the first time that she meets the "famous" Frank Churchill that she has been hearing and dreaming about for years; and this first element of surprise prefigures other, bigger, surprises that this new acquaintance will bring her.

Sometimes, however, the act of opening a door, even in one's own house, marks an intrusion or indiscretion. This is the way Elizabeth Bennet feels when she opens the drawing room door and sees Jane and Mr. Bingley hastily moving away from each other (*P&P* 383). Her indiscretion does not have any negative consequences, though, for Mr. Bingley has already proposed. Symbolically, he and Jane were standing "over the hearth," the heart of a home, when making their first plans for their shared future and that is where Elizabeth sees them. Thus, here again we meet the same configuration of fire and doors as the inside-outside opposition.

In other cases, the sudden opening of the door can be more "dramatic." In *MP*, during the theatricals, and at the very moment when Fanny is under pressure to read Mrs. Grant's part, "the door of the room was thrown open, and Julia appearing at it, with a face all aghast, exclaimed, 'My father is come! He is in the hall at this moment'" (*MP* 202). This event marks an important turning point which will have linear consequences for the story: the constellation of the household will be rearranged as Fanny gains in importance and the other Bertram children, except Edmund, take their distance from Mansfield Park, as I discussed in my first chapter. Moreover, the situation is paradoxical, because Sir Thomas "intrudes" into his own home converted, during his absence, into a theatre, although the performance has not yet taken place. In fact, his unexpected return prevents the final transformation of his ancestral home space in a way he could not approve of.

There is another important instance of the opening of a door as marking Sir Thomas' unwelcome presence in his home space. This time it causes anxiety to Fanny for he opens the door of Fanny's room and "the terror of his former occasional visits to that room seemed all renewed [...] as if he were going to examine her again in French and English" (*MP* 360). This time, however, the purpose of his visit is weightier, for he tries to induce Fanny to marry Mr. Crawford whom she does not love. Noticeably, this is the first time that he enters adult Fanny's room and finds out that she has not got a fire in it. He discovers her "nest of comforts" (*MP* 179) and invades it at the same time as he tries to invade her mind and influence her principles. Later, she as reluctantly opens the door of her room to her rival and false friend Mary Crawford (*MP* 413). The only person that she enjoys seeing there seems to be Edmund even if he enters while she is away and she discovers him sitting there at her table: "such a sight having never occurred before, was almost as wonderful as it was welcome" (*MP* 303). To enter somebody's room during their absence is a mark of intimacy; and even though Edmund still perceives this as a brotherly intimacy at this point, Fanny is happy to have him in her personal space.

Closing a door can be as emotionally significant as opening it. In *NA*, for example, the General renders the process of getting things organized and the carriage loaded so complicated and stressful that once the carriage door is shut with only the girls inside—"Catherine's spirits revive(d)" (*NA* 158–159). In this case, the door has a protective function: it allows Catherine and Eleanor to be together and psychologically at ease in a space that is clearly delimited and protected from outside intrusion. The door plays a similar sheltering role in *S&S* when Elinor learns from Edward that he is not married to Lucy. She ran out of the room and "as soon as the door was closed, burst into tears of joy" (*S&S* 408). Social rules only allow her to give way to her emotions in private, and even in such an extraordinary moment Elinor does not forget to close the door after her. Neither her family nor Edward follow her into the room, for they respect her intimacy. We find a very different case with Charlotte Lucas in *P&P*. Out of self-interest Charlotte defies the right to privacy and secretly stays in the room while Mrs. Bennet discusses with Mr. Collins Elizabeth's refusal to marry him (*P&P* 127). Her eavesdropping allows her to give a completely different direction to her life; thus, by renouncing to leave the drawing room she symbolically passes a threshold of taking her destiny into her own hands.

Emma Woodhouse, on the other hand, engaged in her marital schemes for Harriet, is frustrated at not being able to close the door and give the supposed lovers some intimacy:

> She was obliged to leave the door ajar as she found it; but she fully intended that Mr. Elton should close it. It was not closed, however, it still remained ajar; but by engaging the housekeeper in incessant conversation, she hoped to make it practicable for him to chuse his own subject in the adjoining room. For ten minutes she could hear nothing but herself (*E* 96).

The passage is comical for Mr. Elton seems to foil all her attempts at closing him up with Harriet. His goal is to stay in the same open space with Emma. In general, he manages to remove all the obstacles, and overstep all the thresholds that Emma unconsciously builds between him and herself: the quasi-constant presence of Harriet and various "activities" related to her, Mr. John Knightley in the carriage when going to the Christmas party, and Mr. and Mrs. Weston with their stories about Frank Churchill. At the end of the day—Christmas Eve—he is alone with Emma, with the carriage door "lawfully shut on them" (*E* 139) and he gets an opportunity to make his unwelcome proposal.

For Catherine Morland in *NA*, doors that are not or cannot be opened seem to be hiding something mysterious behind them. She struggles to open the chest and the black cabinet in her room (*NA* 172–173), her imagination wild at work, only to find a laundry list. Nevertheless, she is still suspicious of "several doors neither opened nor explained" (188) and decides to investigate the mystery asking herself "to what might not those doors lead" (*NA* 190). Most embarrassingly, she is discovered by Henry Tilney in his mother's room and all her gothic fantasies crumble, as I discussed earlier.

While Catherine Morland learns to open doors with rational serenity, closed doors and gates acquire a more dramatic and symbolic sense in *MP*. After having walked around Mr. Rushworth's terrace the young people, led by Mary Crawford, symbolically pass through the open garden door into the wilderness, which intensifies the psychological tension between

Mary Crawford and Edmund, as well as between Maria Bertram and Henry Crawford (*MP* 106). However, Maria Bertram wishes to go further: through the locked iron gates into the park (*MP* 114) and without waiting till her fiancée fetches the key, she passes round the edge of the gates with Mr. Crawford following her (*MP* 116). This stepping over the imposed limit expresses Maria's unconscious wish to break her engagement, as well as prefiguring her future moves: the elopement, divorce, and her life closed away from society. Thus, by not waiting for a door to be opened, she closes all other doors. On the other hand, squeezing herself round the gates is an act of a desperate person in a desperate situation. Maria wishes to get away from Mr. Rushworth, but by agreeing to marry him, she demonstrates her wish to get away from her father's home. However, as independence is impossible for a woman at that time (except for heiresses such as Emma), and as, contrary to Fanny Price, she lacks strong moral principles and patience, Maria gets emotionally trapped and rushes into a reckless decision. As Jones points out, after the adultery, announced in newspapers, her "chances of any future relationship are nil [...] whereas Henry is free to re-enter society more or less untouched" (106).

The garden door and the gates into the park appear here as a limit—a threshold—that can or cannot be crossed. However, the outcome of stepping over such a threshold may also be more favourable. For example, Mr. Knightley, in *E*, as an old family friend, successfully passes through the door to Hartfield's garden when he sees love-sick Emma walking there (*E* 462), confesses his feelings to her there and gets a positive response. Nevertheless, the garden here is as emotionally charged as the wilderness in *MP*. The accidental meeting of Elizabeth and Mr. Darcy in Pemberley park, in *P&P*, is also one of these emotionally charged moments that reveal characters' deepest feelings. Elizabeth feels embarrassed as if she was invading the privacy of Mr. Darcy whom she had previously rejected, even though she was rightfully led by the gardener to Pemberley gardens (*P&P* 277). This physical and emotional trespassing, however, leads them towards a better mutual understanding which finally ends up in their marriage.

Evidently, the door also constitutes a limit between inside and outside. Thus, often when a welcome or expected visitor arrives they are met at the door and accompanied to the door when leaving. In *P&P*, for example, Jane Bennet is attended to the door by Mr. Bingley when she leaves Netherfield (*P&P* 34), while Charlotte Collins appears at the door when Elizabeth arrives to Hunsford (*P&P* 176), and Mr. Rushworth in *MP* receives "at the door" his "fair lady" Maria (*MP* 98). In such cases the door is a metonym for the house itself: for instance, Emma Woodhouse drives to Randalls' door to hear that nobody is home (*E* 202), or suddenly decides not to pass Miss Bates' door "without going in" (*E* 165). We also find the perfect illustration of this inside-outside limit when Emma is standing at Ford's door and observing "the traffic" of "the busiest part of Highbury" (*E* 251). Here, the normally inquisitive Emma is in-between: concerned and unconcerned at the same time.

The entrance door is also the last limit of the inside space, the last occasion to look back before going away towards an apprehended future. For example, in *P&P*, Mr. Darcy after having heard the story of Lydia's elopement leaves Elizabeth in her room "with one serious parting look" (*P&P* 307). This look on the threshold is significant for, as the reader and Elizabeth learn later, at that moment he is elaborating a plan how to rescue Lydia from public disgrace which would save for him the possibility to marry her sister. Similarly, in *MP*, William Price, anxious for his sister's well-being in her childhood home, turns back at the door and asks his mother to take care of Fanny (*MP* 449).

In *NA* we have an example of how this limit can be reinforced by a supplementary obstacle. When Catherine and Mrs. Allen arrive at their first public ball, they must push through the men grouped at the ballroom door (*NA* 13). For Catherine it is the first social experience of Bath; moreover, it represents a symbolic passage into the real adult world contrasting with the world of her imagination. The situation is, however, more symbolically dangerous when the door does not constitute an effective obstacle, as we have already seen in the passage about the iron gates in *MP*. In *S&S*, Willoughby does not meet any obstacles when he carries injured Marianne into her house: he passes directly through open garden gates and straight into the drawing room (*S&S* 50). This parallels their coming relationship that will advance speedily, without any regard to social conventions, and, unfortunately for Marianne, end in as abrupt a way. There are other episodes in the novel that show Willoughby as a man who oversteps various thresholds with little hesitation or consideration for social rules and feelings of people concerned by the situation. His visit to Allenham with Marianne is figuratively "doorless," for he takes her there without the knowledge of its owner Mrs. Smith with whom Marianne is not acquainted (*S&S* 79). Moreover, Willoughby flees Allenham once he is asked to take responsibility for his behaviour concerning Eliza who had his child, and, without stumbling over any symbolic doorsteps, in a few months marries a rich lady in order to secure his position. At the end of the day, however, he gets his "well-deserved" moral punishment: he regrets having left Marianne for the rest of his life.

Windows

Like doors, windows mark the limit between outside and inside.[10] I would even call this limit a link, for, contrary to a door that is kept closed most of the time, a window is a permanent opening that connects the interior and the exterior spaces. This allows windows to function as an information channel, a space for mental escape or a cell of privacy in a crowded room.

First, windows are there to inform characters about the weather and visitors—two crucial aspects of everyday life cyclicity. Female characters seem to be observing windows on a regular basis, for they are often said to spot and identify a visitor before he or she rings the doorbell. Thus, in *P&P*, the Miss Bennets first perceive Mr. Bingley from the window when he returns their father's visit (*P&P* 9), and Mrs. Bennet twice announces to the household his arrival in this manner (*P&P* 369, 415). Elizabeth is drawn to the window by the sound of Mr. Darcy's carriage when he comes to see her at Lambton Inn and introduces his sister (*P&P* 287); similarly, Elinor Dashwood, in *S&S*, rushes to the window after having heard the carriage stop at the door in order to identify an early visitor (*S&S* 210). These few extra minutes or seconds between seeing somebody through the window and seeing him/her in person, allow some time for the character to psychologically prepare herself for the meeting by letting the first flow of emotion pass, or by finding an excuse to avoid the encounter. That is the option Marianne Dashwood chooses during her stay in London. After having paced the room "from one window to the other" in vain expectation of Mr. Willoughby, she leaves the room once she realizes, with frustration, that the visitor spotted outside is Colonel Brandon (*S&S* 196). Likewise, Catherine Morland, in *NA*, suspects Eleonor Tilney of resentfully avoiding her and pretending to be out, after having noticed her arrival home in Bath through the window (*NA* 91). Incidentally, in town, the multitude of surrounding windows can also reveal unexpected things. For example, it is through a window that Anne Elliot, in *P*, gets proof of Mr. Elliot's dishonest character as she sees him talking with Mrs. Clay when he was supposed to be out of town (*P* 240–241).

[10] N. Reynolds (2010) gives a brief overview of the symbolic history of windows in art: she notes that the Romantic era's stress on the view through the window symbolizes "a frustrated longing or a desire to escape" (16).

However, windows are not doors or letters, and only vulgar characters use them as such. For example, in *P&P*, the gossiping and unrefined Mrs. Philips is described as "throwing up the parlour window and loudly seconding the invitation" (*P&P* 81) to the officers. Likewise, in *S&S*, the especially vulgar Mrs. Jennings who brings her daughter to visit the Dashwoods, is said to have no patience to wait until the door is opened, and to start shouting her news through the window (*S&S* 122). In *E*, we find a similar scene with spontaneous Miss Bates: she talks loudly to Mr. Knightley through the window while other guests are gathered in her drawing room (*E* 263).

On other occasions windows can function as a real or pretended dreamspace, where characters can project their minds and consciously or unconsciously distance themselves from an uncomfortable situation. Harriet Smith, in *E*, for instance stands at a window in reverie about Mr. Knightley (*E* 444), while Frank Churchill walks to a window to gather his thoughts and courage when he is trying to tell Emma about his secret engagement (*E* 281). Similarly, after having set right the information about Lucy Steele's marriage, Edward Ferrars, in *S&S*, rises and walks to the window (*S&S* 407), while he is waiting for Elinor's reaction to the news which would inform him about her feelings. This is also the case with Catherine Morland, in *NA*, who keeps looking through a window after Henry Tilney has left for his parsonage "in a hope of catching another glimpse of his figure" (*NA* 179); and Elizabeth Bennet, in *P&P*, who keeps walking to the windows in Pemberley to enjoy the views (*P&P* 272) which induce her to step beyond her former prejudice. We have a slightly different, more conscious, example with Fanny Price, in *MP*, when her uncle visits her room and tries to convince her to marry Mr. Crawford. She fixes her eyes on the window while Sir Thomas talks to her (*MP* 362), which expresses her wish to avoid the conversation as well as to escape the marriage that is being forced on to her.

As I mentioned earlier, windows can also help to isolate oneself from other people in the room, to delimit a private corner: that is what sceptical Mr. John Knightley is in the habit of doing when he meets his brother during a dinner party (*E* 316). Often, however, this need occurs when characters are submerged by strong emotions that cannot be shared. That is the reason why Fanny Price, when jealous of Mary Crawford, tries to keep Edmund at the window in her intimate little world as long as possible (*MP* 132). However, he "turns his back on the window" and goes to listen to Mary's playing; thus, Fanny, not wanting to join the others, with her aching heart stays to "sigh alone at the window" (*MP* 133). Likewise, Captain Harville, in *P*, wordlessly invites Anne to join him at the window (*P* 251), when he wants to talk about his dead sister and his disappointment in his friend—the matter that touches him deeply. More comically, in *E*, Mr. Elton and Harriet's standing at a window together (*E* 96) is the confirmation of their being "lovers" in Emma's eyes.

In general, then a door represents a linear passage from one type of space to another, as well as from one type of physical or psychological state to another. A window on the other hand only allows characters to see, to project, to dream, and to hide, in other words, to find a different internal "time," while staying in the same space.

Works Cited

Primary Sources

Austen, Jane. *Northanger Abbey*. Ed. Barbara M. Benedict and Deidre Le Faye. Cambridge: Cambridge University Press, 2006a.

Austen, Jane. *Sense and Sensibility*. Ed. Edward Copeland. Cambridge: Cambridge University Press, 2006b.

Austen, Jane. *Pride and Prejudice*. Ed. Pat Rogers. Cambridge: Cambridge University Press, 2006c.

Austen, Jane. *Mansfield Park*. Ed. John Wiltshire. Cambridge: Cambridge University Press, 2005a.

Austen, Jane. *Emma*. Ed. Richard Cronin and Dorothy McMillan. Cambridge: Cambridge University Press, 2005b.

Austen, Jane. *Persuasion*. Ed. Janet Todd and Antje Blank. Cambridge: Cambridge University Press, 2006d.

Austen, Jane. *Later Manuscripts*. Ed. Janet Todd and Linda Bree. Cambridge: Cambridge University Press, 2008.

Secondary Sources

Bachelard, Gaston. *The Poetics of Space*. Boston: Beacon Press, 1994.

Billi, Mirella. "Spaces as Self: Houses in Jane Austen." *The House of Fiction as the House of Life*. Ed. Francesca Saggini and Anna Enrichetta Soccio. Newcastle: Cambridge Scholars Publishing, 2012. 60–71.

Duckworth, Alistair M. *The Improvement of the Estate*. London: Johns Hopkins, 1971.

Johnston, Freya. "Public and Private Space in Jane Austen." *English: The Journal of the English Association* 46 (1997): 193–212.

Jones, Hazel. *Jane Austen & Marriage*. London: Continuum, 2009.
Le Faye, Deirdre. *Jane Austen. The World of her Novels*. London: Frances Lincoln, 2002.
Mullan, John. *What Matters in Jane Austen?* London: Bloomsbury, 2012.
Pidduck, Julianne "Of windows and country walks: frames of space and movement in 1990s Austen adaptations." *Screen* 39.4 (1998): 381–400.
Purdie, Susan. "Mr Allworthy's 'Good House.'" *The House of Fiction as the House of Life*. Ed. Francesca Saggini and Anna Enrichetta Soccio. Newcastle: Cambridge Scholars Publishing, 2012.10–17.
Reynolds, Nicole. *Building Romanticism: Literature and Architecture in Nineteenth-Century Britain*. Michigan: University of Michigan Press, 2010.
Saggini, Francesca and Anna Enrichetta Soccio, eds. *The House of Fiction as the House of Life: Representations of the House from Richardson to Woolf*. Newcastle: Cambridge Scholars Publishing, 2012.
Skinner, John. "Exploring Space: The Constellations of Mansfield Park." *Eighteenth-Century Fiction* 4.2 (1992): 126–148.
Varey, Simon. *Space and the Eigtheenth-Century English Novel*. Cambridge: Cambridge University Press, 1990.

CHAPTER 4

Spatiotemporal Movement and Psychological Change: Walking and Dancing

The close relationship between the cyclical and the linear in Austen's novels is illustrated again in walking and dancing: in other words, through physical movement in space and time which brings psychological movement, or change. The body that produces any physical movement, according to H. Lefebvre "produces itself in space and [...] also produces that space" (170). Moreover, from his point of view, it does not "fall under the sway of analytic thought and its separation of the cyclical from the linear," but rather "unites cyclical and linear, combining the cycles of time, need and desire with the linearities of gesture, perambulation, prehension and the manipulation of things" (203). In dancing, on the physical level, the body works through rhythm, merging the cyclicity of the melody with the linearity of the beat (205), as well as uniting the recurrence of the choreography with the finitude of the step and the gesture. Likewise, in walking, the cyclicity of a garden pathway, sometimes forming an actual circle, and the rhythmical nature of the act itself merges with the decisiveness of direction and destination.

In all of Austen's novels walking, as an exercise, is a part of everyday social cyclicity, a certain ritual that must be respected. Mr. Woodhouse, for example, takes his well-defined "three turns" in the garden (*E* 60) even when Mr. Knightley comes for a visit. The only cases of *force majeure* that prevent characters from their ritual walks are bad weather and sickness. Walking also allows characters to enter a different, open space that can give a new perspective on a situation as well as providing them with some

R. Baublytė Kaufmann, *The Architecture of Space-Time in the Novels of Jane Austen*, https://doi.org/10.1007/978-3-319-90011-7_4

time for privacy, reflection, and, sometimes, for a linear decision. Dancing, on the other hand, permits the creation of a different space within the usual inside space, and like walking, forms a specific private time cell, which allows a linear change to happen.

Of interest too is what initiates the movement, when and why a character is driven to walk several miles in mud (*P&P* 36), to pace the room "from the window to the fire" (*S&S* 216) or to dance with a man "whom one is determined to hate" (*P&P* 101). As L. Worsley points out, even though a lot of Austen's scenes take place in a drawing room "when Jane's characters want to talk about what really matters—their feelings, the truth—they often have to go outdoors" (1). Thus, emotions often make Austen characters move, although movement—notably, dancing—can also generate emotions. For example, despite the strict social rules concerning the choice of dancing partners, dancing itself often rhymes with happiness in the novels. In contrast, walking frequently serves the purpose of calming agitated feelings and of giving some "airing" to thought. Thus, the characters in the novels are moving physically, socially, and emotionally in cyclical and linear ways that allow them to experience their bodies, their feelings and their places in spatiotemporal reality.

Gender is obviously relevant if we consider the social possibilities of movement, on the one hand, for genteel women, whose mobility was restricted by various conduct rules and educational barriers, as well as by the type of clothes they wore; and, on the other, for men of the same class, who enjoyed active, outgoing, lives at the beginning of the nineteenth century.[1] This inequality between the possibilities for men and women is captured by Henry Tilney when he compares dancing with marriage in *NA:* "man has the advantage of choice, woman only the power of refusal" (*NA* 74). The sociohistorical context will thus be taken into consideration in the following analysis of walking and dancing in the novels.

Walking

Historically, according to J. A. Amato, walking can be divided into two categories: walking by necessity, which distinguishes the lower classes, and walking by choice, which distinguishes the upper classes that have the leisure to stroll and promenade (73). At first sight we seem to find both types of walking in Austen's novels, even though walking for choice is evidently

[1] See D. Le Faye's chapter on female fashion (93).

dominant. The walkers by necessity are two all-female families, made poor by the death of husbands and fathers: the Dashwoods in *S&S* and the Bateses with their niece and grand-daughter Jane Fairfax in *E*. However, both families have their neighbours' carriages or servants available, and they only need to walk if they choose to refuse the offered help. Moreover, both families of women also do a lot of leisure walking, therefore we can assert that Austen's walkers are predominantly walkers by choice.

Another pertinent historical distinction is between walking and travelling. A. D. Wallace sees walking as "an insular and confining act," which takes people on errands and visits but keeps them tied to their home place:

> Walking set boundaries and did not break them, it moved in its own circles and did not move through, it remained at home and did not travel, and the walker, although he might be the guide within the labyrinth, did not ordinarily leave it. […] Before the transport revolution, walking ordinarily was the localizing movement opposed to travel. (26)

Austen's novels perfectly illustrate this opposition, where walking never serves a greater practical purpose than reaching a neighbouring house, whereas travelling is an expedition of one or more days in a carriage.

In addition, walking, in the novels, is often referred to as the principle exercise for females and old, sickly men like Mr. Woodhouse. Men also walk, but often for practical, social, or emotional reasons, which I will discuss later. Very few of Austen's heroines take the other form of exercise, which is riding. The one who does ride for her health is Fanny Price in *MP* (*MP* 43), since walking is too demanding for her (*MP* 85). Jane Bennet in *P&P* rides once, for strategic purpose, while the Miss Dashwoods, who used to be riders, cannot afford to keep a horse anymore. These examples show that riding is closely connected to the financial situation of the household, and as Austen's heroines often have financial constraints they mainly walk.

There are other characters in the novels that prefer carriage rides to walking. A. D. Wallace argues that walking in Austen is a "sign of virtues" (99), and she uses this moral approach to classify the characters. She claims that the characters "are distinguished as readily by their willingness, even their desire, to walk as by their common sense, independence of opinion, unostentatious economy, country living and, what Mr. Knightley calls 'English delicacy towards the feelings of others'" (*E* 99). Thus, admirers of the "barouche-landau" like Miss Bingley, Mr. Collins, Mrs. Elton, John Thorpe, or Elizabeth Elliot do not fall into this category of distinction.

Wallace goes further stating that "the differences between carriage-riders and walkers thus signal not only differences between the values of individuals but between what Austen perceives as the value of the literally mobilized *nouveau riche* and those of the traditional landed culture" (101). According to this approach, walking represents traditional cultural values and is anchored in the social rituals of the landed gentry.

I would like to take a different approach and look at walking as a physical, emotional, and social act touching deep layers of characters' consciousness. I will deal mainly with three heroines—Elizabeth Bennet, Marianne Dashwood, and Emma Woodhouse—for whom walking is important on these three levels. I will start with Elizabeth, who is the most physical walker, then, continue with Marianne, who is the most emotional, and finish with Emma, who is the best example of a social walker. Other Austen heroines walk much less and in different situations, although some of their walks are still significant for the narratives. Adopting this approach thus gives us a different grouping of the six novels than the usual classification of the three earlier and the three later works.

Elizabeth Bennet: Springing and Rambling

One of the most famous scenes of walking in the novels is Elizabeth Bennet's three miles' walk to Netherfield to see her sick sister Jane (*P&P* 35–36). She does not show any reluctance to walk such a distance, on the contrary, once her father starts to hesitate to "send for the horses" (*P&P* 35) she decidedly rejects the offer. Therefore, Elizabeth's walk to Netherfield falls under the category of walking by choice. The following passage conveys the rhythm, the tempo, and the energy of her actual "escapade":

> Elizabeth continued her walk alone, crossing field after field at a quick pace, jumping over stiles and springing over puddles with impatient activity, and finding herself at last within view of the house, with weary ancles, dirty stockings, and a face glowing with the warmth of exercise (*P&P* 36).

Walking is described here as an energetic physical activity through body related verbs: crossing, jumping, springing, (face) glowing. The passage communicates the rapidity of her movement initiated, undoubtedly, by her psychological determination to get to her sister as fast as possible, but also by the capacity of rhythmical movement to generate energy. This "flying" passage and, especially its last line, evoke the pleasure Elizabeth feels in the exercise: she seems to be completely absorbed by the physicality of walking, without there being any place left for thoughts. She cuts through the outside space—"crossing field after field"—without noticing it, only conquering obstacle after obstacle in her internal "impatient" time, until she reaches the destination and gets back to the world of social codes and propriety with her petticoat "six inches deep in mud" (*P&P* 39).

Elizabeth's walk is quite exceptional in pre-twentieth century fiction, and, like the walk itself, she is clearly crossing the borders of propriety while she is crossing the fields. She is shown immersed in a physical activity and enjoying it without the excuse of "airing" of thoughts, which anticipates the twentieth century "discovery" of the pleasure of vigorous exercise.[2] According to Pidduck, "such physical mobility expresses a longed-for pleasure in physical movement for its own sake, something historically problematic for women both physically and in representation" (393). The basic etiquette of the time demanded that an unmarried lady never walk without a companion, except "for a walk to church or a park in the early morning" (Pool, 55). Elizabeth walks straight after breakfast which could be considered as "early morning," but she goes way beyond the limits of the park. Moreover, she manages to overcome the constraint of a long dress with petticoats, which is supposed to restrict all not quite lady-like movements, in order to jump over the stiles and the puddles.

This passage is also a perfect example of fusion between the cyclical and the linear in the act of walking. The walking pace, the recurrent obstacles in the form of stiles and puddles, as well as the succession of fields belong to the cyclical, whereas the clearly determined destination she seeks to attain as fast as possible gives a sense of directness, thus of linearity. On the psychological level, her walk is linear by its intention, for here, she does not engage herself in a social outing and does not "waste" any time in admiring the landscape, but rushes to see her ill sister. Moreover, retrospectively, the outcome of her walk is linear as well: her socially eccentric behaviour together with "the brilliancy which exercise had given to her complexion" (*P&P* 36) deepen Mr. Darcy's interest in her.

[2] Claudia L. Johnson remarks that this passage demonstrates how "Austen's early work sides with the body and its vigors, particularly as these give the lie to restrictive, high-blown notions of female delicacy" (22). Johnson contrasts this example with *Persuasion*, where "the pain rather than the vigor of the body [...] seems to occupy Austen's attention" (22). In this case, we could draw a parallel with the declining health of the author herself. However, her last, unfinished novel *Sanditon*, which she wrote while fatally ill, deals with the severe cases of hypochondria rather than any real feebleness of the body.

This passage is not the only one where we see Elizabeth running. Several pages later, she runs "gaily off" leaving the Bingley sisters and Mr. Darcy to walk alone (*P&P* 58), and in this case, together with her physical spontaneity, the act of running demonstrates her emotional detachment from the socially superior company. Later on, she is said to be "in the habit of running" (*P&P* 333), which would imply that she practices a regular demanding physical exercise.

Furthermore, while staying in Kent, she has her favourite paths and walks there frequently alone (*P&P* 191), which indicates that walking must be an enjoyable activity for her, preferable to socializing with Lady Catherine, the Collinses, and, later, with Mr. Darcy who, annoyingly, every time he meets her walking "[thinks] it necessary to turn back and walk with her" (*P&P* 204). The place she chooses for her favourite walk, significantly "edges" the park, thus it is situated between sociocultural and "wild" spheres: it "was along the open grove which edged that side of the park, where there was a nice sheltered path, which no one seemed to value but herself" (*P&P* 191). This reflects Elizabeth's character, which confronts propriety and social awareness with spontaneity and independence of thought and action. She likes to keep her views open, while staying on a protected and relatively conventional path, unlike her sister Lydia who recklessly throws herself into the socially condemnable act of elopement. This walking space also serves as a refuge from possible unwelcome intrusion, as she feels quite a stranger with the Collinses, Maria Lucas, and Lady Catherine. Moreover, she evidently takes pride in being the only one capable of appreciating the place. For the experienced reader, Mr. Darcy's choice of the same path indicates similarities of taste and anticipates their future sharing of the same space. This at least is the case the first time—later he presumably chooses it on purpose to "accidentally" meet Elizabeth.

In contrast to her purely physical walk to Netherfield the time of Elizabeth's solitary "rambles" in Kent is often dedicated to thinking and re-thinking. First, she is said to take Jane's letters with her on her walks and to "dwell on some passages" (*P&P* 204) trying to infer her sister's real emotional state from between the lines. The cyclicity of the act of walking helps her to concentrate on the same words again and again until they acquire (or not) a new meaning. The open, natural space, as opposed to emotionally and socially charged house space, also helps her to clear her mind of its prejudice and to have a different, "outside" perspective. The cyclicity of the act thus leads to the reaching of a linear understanding. Similarly, later in the novel we see her father Mr. Bennet walking in the copse and reflecting on his brother-in law's letter concerning Lydia's marriage to Wickham (*P&P* 332–333). The outcome of this reflection is reached after he hears the letter read out loud: he manages to conclude that Mr. Gardiner must have paid Wickham at least ten thousand pounds to induce him to marry. Having, thus, made sense of the situation, Mr. Bennet feels ready to turn back and walk home (335). This passage, when compared to the passages on Elizabeth's reflective ramblings discussed above also illustrates the likeness between the father and the daughter.

Nevertheless, it is most frequently by her emotions that Elizabeth is propelled to walk or/and to keep walking. One of the strongest emotional moments occurs the morning after Mr. Darcy's first proposal which she chooses to handle outside with the help of "air and exercise" (*P&P* 217). At first her attention seems to be turned outwards, her mind not being yet open to a deeper analysis of recent events: the narrator describes in detail her decisions—to take one path, to turn, to walk "two or three times along that part of the lane" (*P&P* 217) and, most significantly, her observing of the advancing Spring. However, from the moment when Mr. Darcy gives her his letter, she turns inwards: first to read and then, to reflect and start reforming her earlier perceptions. Elizabeth is "wandering along the lane for two hours, giving way to every variety of thought; re-considering events, determining probabilities, and reconciling herself as well as she could to a change so sudden and so important" (*P&P* 231). The passage suggests that she is pacing the lane back and forward, without noticing the surroundings any more, and at the same time she is going/walking through her memory, readjusting her impressions and remapping the web of relationships. Noticeably, Elizabeth is reading while moving and, in parallel, her attitude towards the author of the letter and the history of their relationship is shifting. By the end of her "wandering" she seems to acknowledge that the change is definite, and it is now up to her to accept it. Thus, during these two hours of internal space/time travel, a linear transformation takes place.

Later, after Mr. Darcy's and Mr. Bingley's visit to Longbourn the following Autumn, Elizabeth escapes again for a walk "to recover her spirits; or in other words, to dwell without interruption on those subjects that must deaden them more" (*P&P* 375). This time her emotional storm is due to not knowing if Mr. Darcy still has feelings for her. Similarly to her walk in Kent, she mentally goes through his behaviour trying to infer his thoughts. The conclusion of the walk is very different though as she has no time to dwell on the subject till it loses its emotional intensity, for she is obliged to put her wonderings aside by the arrival of Jane.

Elizabeth's "discovery tour" in the park of Pemberley can also be described as emotional walking. Although disguised as a walk done out of a tourist's interest (which is what it is for the Gardiners), the walk allows Elizabeth to connect emotionally to Mr. Darcy by experiencing the environment in which he grew up and which he owns. The windings of the coppice-wood that Elizabeth "longed to explore" (*P&P* 280) convey her strengthening desire of closeness with his owner. Her emotional rambling is interrupted by the sudden appearance of Mr. Darcy, which changes her dreamy state into embarrassment and a sense of intrusion. Exploring somebody's garden seems to be comparable to intruding into somebody's intimate space as I have mentioned in Chap. 2. Symbolically, she finishes her walk together with Mr. Darcy who indirectly asks to see her again (*P&P* 283), which signifies that he accepts her in his personal space and even invites her to "explore" further. This accidental meeting, thus, anticipates the decisive development in their relationship.

Elizabeth does not always walk alone. She also walks with those with whom she wants to share intimacy. This is in general the case when Jane and Elizabeth, who are especially close to each other, go for a walk together (*P&P* 332, 375). Likewise, at the end of the novel, Elizabeth and Mr. Darcy, as well as Jane and Mr. Bingley, go on long walks for the same reason: to be together and talk without being heard by the rest of the household. During these walks the lovers can deepen their emotional connection to each other as well as clear up the inevitable misunderstandings of the past (*P&P* 407). In fact, Mr. Darcy's second marriage proposal and its acceptance by Elizabeth take place during their walk in the outskirts of Longbourn. The emotional turbulence that they both experience at that moment makes them lose sense of time and space:

> They walked on, without knowing in what direction. There was too much to be thought, and felt, and said, for attention to any other objects. (*P&P* 407)

Their conversation turns round and around different moments of their history while they are engaged in the cyclicity of walking. The situation is comparable to Elizabeth's solitary ramblings in Kent, but this time it is shared by two people symbolically advancing in the same direction. The linearity of the proposal is, thus, followed by the cyclicity of mutual re-reflection and grounded by the rhythm of steps.

Retrospectively, the reader can notice how this passage is anticipated by a comically similar one in chapter 22 of the first volume, when Mr. Collins proposes to Charlotte Lucas. In contrast to the long and complicated path that leads the real lovers—Elizabeth and Mr. Darcy—to their union, Mr. Collins proposes to Charlotte on the third day after having been refused by Elizabeth. The arrangement takes place while they are walking in the garden of Lucas Lodge, and once they enter the house "everything [is] settled between them to the satisfaction of both" (*P&P* 137). This is a purely linear scene, pointing clearly to Charlotte's[3] economic interest and Mr. Collin's determination to select his wife during this visit no matter what. Both characters are well aware of their motives and head directly to the house to announce the news "in as short time as Mr Collins's long speeches would allow" (*P&P* 137).

As well as walking for exercise or to deal with emotions there are numerous instances of social walking in the novel. For instance, the Miss Bennets regularly go to Meryton together to get some news and, eventually, make new acquaintances (*P&P* 80, 129). The period of rain preceding the Netherfield ball that makes walking impossible for ladies is said to be a difficult time to bear: "no aunt, no officers, no news could be sought after" (98). Thus, walking, in the case of the Bennet girls, allows them to socialize and fill their countryside lives with some excitement. Similarly, we hear that the Collinses, and especially Mr. Collins, walk daily to Rosings (*P&P* 190) to see Lady Catherine. Elizabeth considers these visits as mercenary, thinking that there "might be other family livings to be disposed of" (*P&P* 190). On the other hand, Lady Catherine seems to represent the principal, and, possibly, the only social bond that the Collinses have in Kent: the narrator mentions that "the style of living of the neighbourhood in general, was beyond the Collinses reach" (*P&P* 190).

There are cases of social walking too that serve the purpose of social exclusion. This we see in two walking scenes during Jane and Elizabeth's stay in Netherfield. The first one, is the walk in the park that starts with two pairs—Elizabeth with Mrs. Hurst and Mr. Darcy with Miss Bingley. Once their pathways cross, Mrs. Hurst leaves Elizabeth and takes Darcy's other arm, thus taking all the space on the pathway and pushing Elizabeth out of their group (*P&P* 58). This is a strong physical gesture, reinforced by the symbolic "appropriation" of Mr. Darcy signalling that there is no place for Elizabeth in their social world. However, Elizabeth's decision

[3] Charlotte must have felt pressure from her family to accept any eligible marriage proposal she could get, for she was 27 years old and not a good looking woman. This is because at this time women's "economic security lay in male hands—fathers', brothers', husbands'" (H. Jones, 2). In the novel we learn that Charlotte's brothers "were relieved from their apprehension of Charlotte's dying an old maid" (137), because after the death of the father, her brothers would have had to take care of her.

to run "gaily off" (*P&P* 58) instead of clinging to social superiors after Mr. Darcy politely proposes that they "go into the avenue" (*P&P* 58), ironically, boosts his interest in her.

The other walking scene at Netherfield, curiously takes place inside, in the drawing room, when Caroline Bingley asks Elizabeth to "take a turn about the room" under the pretext of it being "very refreshing after sitting so long in one attitude" (*P&P* 61). Evidently, the motive of exercise is here a cover for Caroline's husband-hunting strategy. Her real purpose, as Mr. Darcy works out a minute later, is to display her figure, "a certain something in her air" and her "manner of walking" which she considers as necessary accomplishments for a lady (*P&P* 43) and which, she visibly thinks, Elizabeth is lacking. Thus, by walking she seeks to mark the social division between her and her rival.

In addition, and quite surprisingly, a way of walking can recommend a man, too. When Elizabeth notices Mr. Wickham for the first time, he is walking and she, together with her sisters, are "struck with the stranger's air" (*P&P* 80). Later, during the evening party at Mrs. Philips house, Elizabeth sees Mr. Wickham as "far beyond them all [the other officers] in person, countenance, air, and walk" (*P&P* 85). However, in the end, Mr. Wickham turns out to be as falsely accomplished as Caroline Bingley. Thus, an excessive attention to a manner of walking as an apparent sign of superiority is a mark of superficiality and lack of real moral discernment.

Wandering Marianne

Marianne Dashwood in *S&S* is also a famous Austen walker. Like Elizabeth Bennet often engaged in her independent "rambles" (*P&P* 204), Marianne is a solitary walker: she is regularly "wandering alone" (*S&S* 32, 96, 343) driven by her emotions. The very first example of her walking takes place during the last day at Norland. Marianne is wandering "alone before the house" (*S&S* 32) deeply entangled in her melancholic feelings. Walking within the view of her childhood home sharpens the sensation of loss, and Marianne, who believes for the moment that only strong and uncontrollable emotions are worth experiencing, is dwelling in a psychologically trying zone. Unlike Elizabeth Bennet, she does not go for a walk to calm herself or to reflect on a situation in order to understand it better; Marianne chooses to remain in the very heart of her emotion and maintain it at its highest level of intensity as long as possible. The trajectory that she picks for her walk is, thus, calculated to constantly remind her of the most painful aspect of the situation.

During her solitary wanderings, Marianne lingers in the past. When walking in front of Norland house, she is mentally going over the childhood years passed in that space, and even though, in her earlier quoted soliloquy, she evokes her departure, she stresses the fact that the trees will remain and "continue the same" (*S&S* 32), exactly in the same place where they used to be during her life there. Norland, in her perception, becomes an enclosed spatiotemporal cell to which she can pay an emotional visit.

Several weeks later, Marianne again goes through an equally intense experience, after Willoughby's sudden departure. The narrator ironically describes her state of mind right after this event:

> Marianne would have thought herself very inexcusable had she been able to sleep at all the first night after parting from Willoughby. She would have been ashamed to look her family in the face the next morning, had she not risen from her bed in more need of repose than when she lay down in it. [...] When breakfast was over she walked out by herself, and wandered about the village of Allenham, indulging the recollection of past enjoyment and crying over the present reverse. (*S&S* 96)

The passage shows that Marianne takes pride in physical exhaustion caused by emotional suffering, as if embodying the emotion and making it outwardly visible would testify to its authenticity and strength. As in the previous passage, she is wandering alone in the places that remind her of a happy past in order to enhance the force of her present emotions, instead of trying to calm and control them, which would normally be expected from a lady.[4] Moreover, Allenham, the village bordering the estate that Willoughby is supposed to inherit and where Marianne chooses to walk and cry, does not only represent the space of good memories created during their "illegal" visit to the house. The estate of Allenham is also her "dream house," as I discussed in the previous chapter, and as such, it represents her potential future life with Willoughby.

Marianne's emotional wanderings have a cyclical quality. She repetitively engages in the same type of procedure: walking in physical and emotional circles. However, these painful circles gradually, and in spite of her will, become less severely hurtful with only short moments of crises, as her feelings little by little lose their intensity: "such violence of affliction [...]

[4] See Le Faye, 89.

sunk within a few days into a calmer melancholy; but these employments, to which she daily recurred, her solitary walks and silent meditations, still produced occasional effusions of sorrow as lively as ever" (*S&S* 97). Thus, these walks and meditations act like an unsought therapy and bring Marianne back to a temporary emotional equilibrium.

However, Marianne's solitary walks in Cleveland, the Palmers' home, acquire linear quality and have much more serious physical and emotional consequences. Though she has never visited the estate before, it has an imaginary emotional charge for her, because it is relatively close to Willoughby's home—Combe Magna—another place Marianne has never seen, but only heard of and imagined, as I discussed in Chap. 2. Thus, she does not walk to explore the park of Cleveland or to admire the natural surroundings. She is wandering on her internal pathways which all lead in the same direction—Willoughby, whose image still saturates her mind and swells her heart "with emotion" (*S&S* 343).

On the other hand, Marianne takes advantage of the freedom that the countryside, as opposed to London, grants to a woman—the possibility to walk alone, without severely violating the rules of propriety. After having spent several weeks closed up in her guest-bedroom in London with her emotions similarly compressed,[5] Marianne is consciously scattering herself physically and psychologically without any restriction in this open and emotionally charged space until she is exhausted:

> In such moments of precious, of invaluable misery, she rejoiced in tears of agony to be at Cleveland; and as she returned by a different circuit to the house, feeling all the happy privilege of country liberty, of wandering from place to place in free and luxurious solitude, she resolved to spend almost every hour of every day while she remained with the Palmers, in the indulgence of such solitary rambles. (*S&S* 343)

The passage is striking for the oxymoronic phrases: "precious … invaluable misery," "rejoiced in tears of agony," feeling miserable in happy liberty and "luxurious solitude." These combinations of opposites allow the reader to

[5] According to Tanner, Marianne is torn between, what Foucault calls "the absolute exteriority of other people" and "the irreversible interiority of passion and unfulfilled desire" (84). She falls ill from not being able to "join in the social masquerade" (84) and find a sociably acceptable "outward form." As a contrast to Marianne, he points out Elinor's exceptional ability to cover screens, seeing them as a symbolic device to "preserve society" from private feelings, thus, as a selfless act (86).

fully measure the quality of Marianne's personality: she is letting herself plunge into her suffering and is feeling some kind of pleasure in it. Moreover, what looks like a hyperbole at first—"every hour of every day"—turns out to be Marianne's real plan that she is determined to maintain regardless of the wet weather and of any possible social claims. She exposes herself to extreme physical effort by taking "delightful twilight walks [...] all over the grounds, and especially in the most distant parts of them [...] where the trees were the oldest, and the grass was the longest and wettest" (*S&S* 346), as if she unconsciously sought to augment her emotional pain by physical suffering. The superlatives used to describe the space of her roaming underline her emotional state. Thus, her painful and dangerous wanderings at Cleveland, though repetitive, are linear, directed, as she later acknowledges herself (*S&S* 391), to self-destruction.

Marianne falls ill partly because of physically exhausting walks and "sitting in her wet shoes and stockings" (*S&S* 346), but her plan for mental and physical recovery at Barton also contains long daily walks (*S&S* 388). She imagines the directions her different walks will take, which this time have more of a social character, consisting as they do in visits to her neighbours, in contrast with her earlier emotional wanderings (*S&S* 388). Moreover, she has started to associate such social walking with happiness. The first post-illness walk she takes with Elinor is already very different: it is a reflective excursion through recent past events. By "walking" back in time, Marianne gives Elinor her own evaluation of her passions and her behaviour, while Elinor communicates the account of Willoughby's life that she heard from him at Cleveland. Thus, this walk is part of the cyclical healing process through which Marianne will go during the next two years, until her marriage to Colonel Brandon and the beginning of her new life cycle.

Elinor is also a walker, though in contrast to Marianne, she is never shown walking alone in the narrative. Nevertheless, like her sister, she engages in emotional walking, just in a less exposed or unconventional manner. We can see this in the passage quoted in Chap. 2, when Elinor, after the evidence of Willoughby's betrayal, paces the room from window to fire (*S&S* 216). Contrary to Marianne, who chooses wilderness in order to release her spontaneous and stormy emotions, Elinor, whose aim, in general, is to channel and control her feelings, stays inside the house and, thus, manages to dominate her agitation by repetitive, rhythmical walking. Her serene and wise response to Marianne's impulsive demand to go home the very next day (*S&S* 217) proves the success of her "walking tactics."

An important episode of emotional and social walking occurs during Lucy Steele's visit to Barton Park. Being aware of Edward Ferrar's feelings for Elinor, Lucy wishes to inform Elinor of her engagement to Edward in order to limit the risk of his breaking the bond that is no longer based on love. Elinor chooses to walk while extracting the details concerning this past event, for it helps her keep her emotions levelled and allows her, by exposing only her profile to Lucy, to conceal her shock and vexation (*S&S* 148–155). Here, once again, we can see the merging of the cyclical and the linear: a steady pace grounds a linear message that threatens to deprive Elinor of her dream-life.

Most of the moments of social walking in *S&S* take place in London, where the ladies never seem to walk for physical exercise or privacy. London walking is more concerned with watching people and looking at shops. For example, several times in the novel Bond Street is mentioned where "much of their business lay" (*S&S* 187) and where Elinor could not manage to interest Marianne in any of their purchases because Marianne's "eyes were in constant inquiry" for Willoughby (*S&S* 187). Thus, London walking is directed to the surrounding world and destined to bring about meetings with old acquaintances as well as, possibly, making new ones (*S&S* 187, 307).

(Un)accompanied Emma

With the marriage of Emma's governess Miss Taylor in the beginning of the novel, we find the heroine in urgent need of "a walking companion" (*E* 25):

> In that respect Mrs. Weston's loss had been important. Her father never went beyond the shrubbery, where two divisions of the grounds sufficed him for his long walk, or his short, as the year varied; and since Mrs. Weston's marriage her exercise had been too much confined. She had ventured once alone to Randalls, but it was not pleasant. (*E* 25)

Emma is not Marianne Dashwood or Elizabeth Bennet who both enjoy walking alone, but a social walker *par excellence*, who uses walks to talk and scheme. On the other hand, walking is presented here as a necessary part of her daily activities, something, moreover, that requires effort, as we can see from the fact that Emma cannot be satisfied with her father's slow rounds in the shrubbery. It is in part for these two practical reasons that Emma selects Harriet Smith "whom she could summon at any time to a walk" (*E* 25) as her companion. We find the two young women walking together and experiencing various social "adventures," which illustrates the conjunction between the cyclicity of walking and the linearity of an event that happens during a walk. For example, they accidently meet Robert Martin, Harriet's admirer, which allows Emma to adjust her preformed opinion and meddle with Harriet's feelings about him (*E* 31–34); also they are spotted through a window and joined by Mr. Elton, which gives Emma the opportunity to promote her plot for an attachment between him and Harriet (*E* 93).

Highbury and its people make up Emma's living space, her world, every part of which is reachable on foot. Except for Mr. Woodhouse and his physical and/or psychological incapacity to walk further than his shrubbery, everything in Highbury seems to be at a walking distance and its inhabitants are regularly seen walking: Miss Bates and Jane Fairfax go everywhere on foot, Frank Churchill ventures with his father and his stepmother to Hartfield and to the village (*E* 210, 211), John Knightley walks with his sons while visiting the Woodhouses (*E* 117), Mr. Knightley is said to walk to Hartfield almost every day (8) and even the pompous Mr. Elton goes on foot to Hartfield or Donwell (*E* 87, 499). Walking manifestly appears thus as opposed to travelling enhancing the sense of the self-sufficiency of the place as well as its inhabitants' strong connection to it and to each other. All that is beyond walking distance seems far and getting there requires a special preparation, as is the case with the excursion to Box Hill (*E* 399). However, confinement to space within walking distance, is, noticeably, the lot of females. Some men—Mr. Knightley, Frank Churchill, and Mr. Weston—have horses and ride, whenever they need or want to. It might seem surprising that Emma, who is definitely rich enough to keep a horse, does not ride. The nature of the act of riding, however, leaves less space for "accidental" meetings and conversation, which is what Emma seeks for during her social promenades. Walking thus suits her better than riding.

In the last volume of the novel the nature of the walking changes as it starts generating strong and contrasting emotions. First, Harriet's outing with her school friend on the road "deeply shaded by elms" is disturbed by the Gypsy children asking for money (*E* 360). Harriet falls down and is totally incapable of defending herself, like a medieval maiden in distress, while Frank Churchill is given the opportunity to rescue her. As the reader witnesses the scene through Emma's imagination who sees Frank's intervention as a romantic life-saving act, it serves as another misleading narrative device. On the other hand, it also slightly shifts the picture of "security," calm, and isolation from the outside world that characterizes Highbury.

Later, Mr. Knightley's disapproval of Emma's rudeness to Miss Bates on Box Hill shakes her usual emotional balance. Jane Fairfax, on her side, is hurt by the behaviour of Frank Churchill and, as the reader finds out later, decides to break their secret engagement. Thus, we meet both ladies walking on their own. Jane, like Marianne Dashwood, is "wandering about the meadows" (*E* 426), emotionally upset, while Emma, who feels guilty, tries to make amends to Miss Bates and Jane and is obliged to walk alone in "pensive meditations" (*E* 419) when her help is not accepted. Emma's walking alone is unusual for her and it signals that the emotional tension is too strong to be easily shared.

The culmination of Emma's emotional "walking" is her anxious turn in the shrubbery after she has realized the nature of her feelings for Mr. Knightley (*E* 462). It is the first time in the novel that we see Emma looking at nature and seeking for emotional comfort from it:

> Never had the exquisite sight, smell, sensation of nature, tranquil, warm, and brilliant after a storm, been more attractive to her. She longed for the serenity they might gradually introduce; and [...] she lost no time in hurrying into the shrubbery.—There, with spirits freshened, and thoughts a little relieved, she had taken a few turns... (*E* 462)

Emma is shown here as psychologically transformed. After having understood her own feelings and felt the pain of, possibly, unreciprocated love, she has allowed herself to open up her more psychologically vulnerable side and has gone to look for comfort in more "eternal" sources than superficial social interaction.

Emma's "enlightened" state is of short duration for Mr. Knightley unexpectedly returns. As in *P&P*, Mr. Knightley proposes during a walk. However, here it happens in a cyclical space of the shrubbery, close to Emma's home, which suggests her continued life there. Nevertheless, once they come back into Hartfield house their lives are transformed, and all the familiar space appears altered because of her altered state of mind:

> What totally different feelings did Emma take back into the house from what she had brought out! [...] They sat down to tea—the same party round the same table—how often it had been collected!—and how often had her eyes fallen on the same shrubs in the lawn, and observed the same beautiful effect of the western sun!—But never in such a state of spirits, never in anything like it. (*E* 473)

The passage shows that Emma lives an extended moment, an experience of Bachelard's vertical time of a poetical instance (*E* 511) as opposed to the horizontal time of social situations. Her home space with all its habitual features acquires a new, and unique, quality. The moment unites the cyclicity of a daily tea-drinking ritual and the linearity of the life-changing decision, which is still kept secret between her and her fiancé. In effect we might say that, when going out to the shrubbery, she walks out of her childhood home, and, half an hour later, she returns into the same house, transformed into her adult home.

Catherine, Anne, Fanny, and Other Occasional Walkers

The remaining three heroines—Catherine Morland, Anne Elliot, and Fanny Price—are less determined walkers. Fanny is, actually, hardly a walker at all, while Anne, who seems to possess all the qualities of a solitary rambler is never shown walking alone except for practical purposes. However, in all three novels we find some important examples of walking that have significant linear consequences.

Wallace's opposition between rational as well as aesthetically educated walkers and superficial carriage riders in *Northanger Abbey* (101) is illustrated by Catherine Morland's being torn between these two forms of motion which are associated respectively with two very different men who are competing for her attention. The carriage rider John Thorpe tricks her into joining him in his hardly proper and perilous ride, because of which she misses the walk with Henry Tilney and his sister (*NA* 78). The ride with Thorpe is presented as an unpleasant and stressful "brisk" escapade into which Catherine is "whisked" with "no power of getting away" (*NA* 86). Moreover, the promised visit to Clifton, Kingsweston, Blaize Castle, and "anything else we can hear of" (*NA* 83) as Mr. Thorpe describes it, never takes place. It turns out to be a pointless ride that upsets Catherine's emotional balance. We still feel this hectic mood two chapters later, when the same scheme is imposed on Catherine once again, and her walk with the Tilneys postponed behind her back (*NA* 100). The dispute between Catherine and the Thorpes with her brother James takes a physical turn:

> "This will not do," said Catherine; "I cannot submit to this. I must run after Miss Tilney directly and set her right."

> Isabella, however, caught hold of one hand; Thorpe of the other; and remonstrances poured in from all three. [...]
>
> "Then I will go after them," said Catherine, "wherever they are I will go after them." [...] And with these words she broke away and hurried off. Thorpe would have darted after her, but Morland withheld him. (*NA* 101)

Here we not only see Catherine's obstinate character, her sense of what is morally right and further proof of her strong feelings for Henry Tilney, but we also witness again her spontaneous physicality, her direct relationship to her body that seemed to have been tamed and forgotten once she was a properly behaved lady. From this passage re-emerges the young Catherine of the first pages of the novel, who loves playing cricket and rolling down the slope at the back of the house.

The walk with the Tilneys constitutes a contrast with the ride with John Thorpe in both physical and emotional terms. Henry Tilney takes her to the landscape of his choice and makes her discover it at a walking pace (*NA* 107–116); at the same time as the conversation flows, he opens a different intellectual landscape for her: talking of literature, history, politics, and the picturesque—something far away from Mrs. Allen's talk of dresses and Miss Thorpe's talk of men. Catherine is as receptive and open to Henry's opinions as to the landscape they are viewing from the top of the hill although her inclination to hyperbole is still (comically) present: "Catherine was so hopeful a scholar, that when they gained the top of Beechen Cliff, she voluntarily rejected the whole city of Bath, as unworthy to make part of a landscape" (*NA* 113). In general, walking with Henry Tilney means an intellectual wandering from one subject to another. Catherine's emotions are positively engaged in following and admiring her mentor, and her self-consciousness motivates her to learn with naive eagerness. The fact of being away from town and society, in a more natural environment, gives enough space and time for such experience. The trip with John Thorpe stands as a contrast in being much more physical, speedy, impatient, and imposed. The conversation conducted by Thorpe turns, moreover, only around money, drinking, and the outstanding performance of his horses (*NA* 59–61).

Henry Tilney also drives Catherine in his carriage. However, this is a very different ride from the one with John Thorpe. First of all, it has a different goal—that of travelling, and not of pure pleasure, thus, the use of a carriage is justified by necessity. Second, Catherine's feelings towards Henry are already strong, as we can see from the exclamation that "to be driven by him, next to being dancing with him, was certainly the greatest happiness in the world" (*NA* 160). And, third, Henry's manner of driving is very different from that of his former rival: "Henry drove so well,—so quietly—without making any disturbance, without parading to her, or swearing at them [his horses]" (*NA* 160). Thus, driving with him could almost pass for walking with him. Their conversation during the journey indicates, however, a greater closeness between them than during the Bath walk: we see Henry pulling all the right strings to tease Catherine and inflame her imagination, by improvising a gothic-inspired story about the "mysteries" of Northanger Abbey. This ride constitutes indeed a transitional moment on different levels: it obviously links the two main parts of the narrative, but it also marks a physical and psychological transition from a socially busy and exposed Bath to the lonely space of the Abbey, which will transform Catherine's imaginary world by breaking her gothic fantasies.

The relationship between Catherine and Henry Tilney, deepened by their country walk in Bath, their ride to Northanger Abbey and the time spent there, as well as at Henry's parsonage, undergoes its principal linear change during another walk, several weeks later. Like Mr. Darcy to Elizabeth and Mr. Knightley to Emma, Henry Tilney proposes to Catherine during a walk—a very short one in this case—from Fullerton Parsonage to the house of the Allens, which is the only moment of privacy that they can get: "they began their walk, […] but his first purpose was to explain himself, and before they reached Mr. Allen's grounds he had done it so well, that Catherine […] was assured of his affection" (*NA* 252). Thus, this is a walk planned in advance in order to speak to Catherine alone. Henry Tilney goes all the way to Fullerton expecting to have this short walk with Catherine, which brings a satisfactory resolution of their abruptly interrupted relationship. On the other hand, it is symbolically important that Henry comes to see Catherine in her home space, and goes for a walk with her there. This allows him to really enter into the space where Catherine grew up and, thus, become part of her life.

In general, as the examples given in this chapter show, the main male characters visit the heroines' homes and walk with them there. The marriage proposals are uttered during the walks (or just after, as in the case of Edward Ferrars) close to the female home space. The symbolic dimension of a man coming to take his chosen woman from her home is thus quite evident in the novels.[6] The only proposal made far from home space—Mr. Darcy's first offer to Elizabeth—is rejected.

[6] According to Franco Moretti, the fact that in her novels, Austen often marries people coming from different counties is significant and new.

The exception to this pattern is *P* in which Captain Wentworth proposes to Anne by letter (*P* 258) in an inn where her relatives are staying while in Bath; however, he gets the answer outside, once he catches up with Anne, who is heading to her new home in Bath accompanied by her brother-in-law Charles (*P* 260). Conveniently and symbolically Charles offers Frederic Wentworth's arm to Anne, and asks him to walk her to her "father's door" (*P* 261). Like Elizabeth with Mr. Darcy, and Emma with Mr. Knightley, Anne and Captain Wentworth lose the sense of time and space while physically walking together and, at the same time, going over their past experiences, vexations, and hopes for the future: "as they slowly paced the gradual ascent, heedless of every group around them [...] they could indulge in those retrospections and acknowledgments, and especially in those explanations of what had directly preceded the present moment" (*P* 261–262). As we have seen in other similar cases, the rhythm of their steps and the cyclicity of a path that they take without consciously choosing any direction merge with the linearity of their decision and help to anchor their excitement.

In contrast again to other novels, where characters mostly walk alone or in pairs, *P* features two important moments of collective walking: a walk to Winthrop (*P* 89–98) and walks in Lyme (*P* 103–118). These are instances of social walking and their principal motor is the interest of the females—the Miss Musgroves, Mrs. Musgrove, and Anne—in Captain Wentworth as well as his interest in them. The walks provide opportunities for exchanges of words, glances, and signs of attention, as well as for overhearing, so convenient for the advancement of the narrative. On the other hand, the physical, bodily dimension of walking is also present. During the first, we see an opposition between two young, energetic girls, Louisa and Henrietta, and two slightly older women, Mary and Anne, whose health and stamina are in question. Louisa and Henrietta suggest a "*long* walk" (*P* 89) hoping perhaps that the hypochondriac Mary and the fragile Anne will not join them, and thus they will not have to compete for Captain Wentworth's attention. However, Mary's wish to socialize turns out to be greater than

> It means that these novels try to represent what social historians refer to as "the National Marriage Market": a mechanism that crystallized in the course of the eighteenth century, which demands of human beings (and especially of women) a new mobility: physical, and even more so, *spiritual* mobility. (15)

Thus, they should be able to feel the "nation-state" as "a true home-land" (15). However, Moretti is missing a step in this process: women are fetched by men from their original home.

her "weak" health. After having reached their destination, the party breaks up in order to pursue their real interests: Henrietta, wishing to make up with her fiancé, and Charles, run to visit Winthrop; Louisa continues walking around with Captain Wentworth looking for hazelnuts which gives them an excuse for intimacy; while the situational "outsiders" Mary and Anne find places to rest on the hill (*P* 92). Although Mary is fidgeting between resting and seeking to be at the centre of attention, Anne is said to be "really tired" and "glad to sit down" (*P* 93). Her overhearing of Louisa's and Captain Wentworth's conversation which includes opinions of her, gives emotional pain, which, together with physical tiredness renders difficult the journey back to Uppercross. Anne is physically and emotionally saved by Captain Wentworth's gesture of handing her into his sister's carriage: "she was in the carriage, and felt that he had placed her there, that his will and his hands had done it, that she owed it to his perception of her fatigue, and his resolution to give her rest" (*P*98). This passage reveals the close connection between physical and psychological health: endlessly sick Mary is given energy by a socially exciting activity, whereas emotionally exhausted Anne suffers an unsurmountable physical tiredness. During the walking scene in Lyme, after Louisa's fall (*P* 118), we see a reversal of the situation: Anne's physical fragility is surpassed by her mental force and ability to act sensibly in a dangerous situation. Thus, Anne's physical strength is again shown to be dependent on her emotional state, and later in the novel, when her relationship with Captain Wentworth is positively advancing, we see her change from a tired, agonizing, and depressive "quasi-spinster" to a strong, determined, and happy woman.

Surprisingly, we do not really see Anne wander alone in the novel. Except a few practical, straight-line walks to her sister's house or to an inn at Bath, she does not go out to ramble in nature like Elizabeth Bennet and Marianne Dashwood, even though she gives the impression of having a solitary walker's disposition. Anne is independent and contemplative, but she does not often seek the rhythm of walking to help her thoughts flow or to deal with overwhelming emotions. There is, however, one telling instance at the beginning of the narrative, when Anne realizes the connection between the Crofts and her rejected suitor:

> Anne, who had been a most attentive listener to the whole, left the room, to seek the comfort of cool air for her flushed cheeks; and as she walked along a favourite grove, said, with a gentle sigh, "a few months more, and *he*, perhaps, may be walking here." (*P* 27)

This passage shows that Anne is familiar with the soothing effect on agitated emotions of walking outside—an effect emphasized by alliteration (the "comfort of cool air")—and that she habitually walks (although it is not mentioned) since she has a "*favourite* grove" (emphasis mine). Her projection into a possible future walk by Wentworth in her "favourite" place betrays, moreover, her belief in the similarity of their tastes and choices, as well as finding a virtual, imaginative way to communicate with him. Thus, not only does she walk to calm her spirits, but she also walks to day-dream. Tellingly, this walk takes place within the limits of Kellynch domain—in her childhood home space. Anne apparently does not venture into spaces with which she is not familiar. Arguably, after having lost her childhood home, Anne does not take solitary walks in other people's gardens and/or public parks seeking emotional comfort, which would be the reason why we do not see her walk alone later in the narrative.

Fanny Price, in *MP*, is the only heroine who does not voluntarily walk. Walking is too physically demanding for her, as we can see from the episode, where Mrs Norris makes her walk on errands for her which gives Fanny a headache (*MP* 85). Moreover, during the walk in the wilderness of Sotherton, discussed in the previous chapter, Fanny spends most of the time seated (*MP* 106); and towards the end of the novel, she is said to reach mutual understanding with Edmund after "sitting under trees" (*MP* 535) with him. However, Fanny rides and that serves as an exercise that is supposed to strengthen her health instead of walking (*MP* 82). The pleasure that Fanny might feel when riding is not acknowledged by other characters in the novel, and her riding time becomes a lonely time, especially after the arrival of the Crawfords. Other young people go on riding expeditions together, making it one of their social rituals. Fanny gets socially excluded, as Mary Crawford takes Fanny's horse.

Thus, walking in all the novels is connected with exercise, rhythm, direction, thinking, private conversation, and social interaction in an outside space. Walking is more prominent in *S&S*, *P&P*, and *E*, where it provokes and develops linear events. In the other three novels walking is less present, however, it is still important for the advancement of the plot. At the level of character, the dreaming Catherine and two rather introverted heroines, Fanny and Anne, walk less than Marianne, Elizabeth, and Emma, who more readily exteriorize their emotions.

Dancing

All Austen's heroines, except, once again, Anne Elliot, dance and enjoy a prospect of a ball, even though some of the balls turn into psychologically injuring experiences. In Austen's England a ball was a strictly ritualized activity with a set of rules concerning notably the order of the dances, how to invite somebody to dance and how to accept or refuse the invitation.[7] The primary social object of dancing was to make young marriageable people meet, have a semi-private moment together "so that they could discover whether their dancing steps and other permissible modes of defining themselves were to fit each other in reasonable comfort" (Sulloway, 144).

Dancing was also a gendered activity with clearly determined roles for males and females. As such, it also served several different functions: social, physical, psychological, and symbolical. For example, the marriageability of a girl was signalled through her first dance at an "official" ball. Judith Lynne Hanna comments:

> Dancing communicates symbolic gender meaning, punctuates phrases of the acquisition of gender knowledge, instills discipline, develops women's muscles that aid in childbirth, builds up endurance, provides recreation, and reflects the personality and physical talents of the individual. Dance may metaphorically enact status transformation from childhood to adulthood. (75)

The "coming out" balls are based on these solid and multi-layered social purposes. We can see in *MP* how important Fanny's first ball is for her, and how interested the men of her family are to see her dance (*MP* 291)! Fanny has the obligation to open the ball, in order to accomplish this symbolic passage into adulthood, even though such a task is much more frightening than pleasurable for her (*MP* 320).

In general, dancing was an important sociocultural activity at the time. In her book *Literature and Dance in Nineteenth-Century Britain*, Cheryl A. Wilson explains:

> For the middle and upper classes, nineteenth-century culture was a culture of dance. Balls and assemblies formed the center of many social circles, both rural and urban [...]. These events played multiple essential roles within communities, from promoting courtship to cementing political alliances. To fully participate in the social and cultural world of the nineteenth century was to dance or, at the very least, to watch others do so. (2–3)

[7] See Jones 45–49 and Pool 78–81.

Wilson illustrates her point with many literary examples, notably Catherine Morland's dancing at five public balls in Bath. She claims that Austen employs the ball scenes to "mark Catherine's steady progress up the social hierarchy through both her relationship with Henry and her education in the codes of polite society" (79). Similarly, we notice that dancing helps to advance relationships between other Austen heroines and socially superior men: Jane and Mr. Bingley, Emma and Mr. Knightley, and, presumably, Emma Watson and Mr. Howard.

Being highly codified and taught by dancing masters, dancing was considered an important female (and male) accomplishment[8] and "one of the few activities that Regency men and women could enjoy together" (Fullerton, 11). Thus, the ability to dance well indicated people's social standing or, at least, their social ambition. Physical pleasure and emotional excitement were not regarded as the goal of a ball. For example, a famous dance master Thomas Wilson asserted that "physical passions had to be subdued and 'mutual purity of conduct' aimed for instead" (Fullerton, 130). Hence, we see that at the time the sociocultural aspects took priority over the physical and emotional effects of dancing. Nevertheless, in Austen's novels, dancing is often presented as an activity to be enjoyed and, as Sulloway puts it, "a classical example of Austenian joy" (138).

Besides the social, cultural, and emotional implications, dancing was a demanding physical exercise, and in order to dance through a whole evening, a young person, especially, a young woman who had few occasions to really move, had to be in good physical shape. Concerning this point, Fullerton notes that "horse riding hardly made a girl breathless, and walking at anything more than a gentle pace was frowned upon, but a ball could provide a good workout for the night" (21). In *MP*, we see that the physically fragile Fanny is seen by her uncle to "rather walk than dance down the shortening set, breathless and with her hand at her side" (*MP* 324). Bearing in mind that this is Fanny's first official ball, it seems hardly surprising. However, more physically active girls, like Elizabeth Bennet, Emma Woodhouse, or ex-tomboy Catherine Morland never seem to be too tired to dance. Good physical shape was also required of a male dancer. In *S&S*, when Marianne, Elinor, and their mother are questioning Sir John about Willoughby, Marianne is impressed by the fact that he could dance "from eight o'clock till four, without once sitting down" (*S&S* 53): she declares that it is "what a young man ought to be"—eager in his pursuits without moderation or fatigue. Noticeably, for Marianne, the

[8] See Fullerton, 15–16.

capacity to dance non-stop for eight hours is not only evidence of good physical shape, but also the mark of a "right" psychological disposition and exciting personality.

Jane Austen in her youth was also a very determined and physically strong dancer, as we can see from her letters to Cassandra Austen. In 1798, when 23 years old, she proudly recounts:

> There were twenty Dances & I danced them all, & without any fatigue.—I was glad to find myself capable of dancing so much & with so much satisfaction as I did. (29–30)

The quotation shows that dancing and satisfaction went together for young Austen; moreover, the phrase "find myself" suggests that dancing permits self-discovery in physical as well as psychological terms. A couple of lines further, in the same letter, she adds, sounding almost as zealous as Marianne Dashwood, that she "could just as well dance for a week together as for half an hour" (30), which proves how much she enjoyed this physically challenging and emotionally engaging activity. In general, the descriptions of balls take a lot of place in her early letters (1796–1800), thus we can conclude that balls and dancing were as important to her in her youth as they are to (some of) her heroines.

In the novels, dancing scenes are not equally numerous and detailed. *NA* and *W* give the most complete descriptions of balls. There are several dancing occasions, presented directly or indirectly, in *P&P*, *MP*, and *E*; however, we find little dancing in *S&S* and *P*, while it is completely absent from the unfinished *S*. Dancing scenes appear notably during the initial period of a relationship between the heroine and her potential husband or they correspond to the beginning of a qualitative change in that relationship, as for Fanny Price with Edmund Bertram and Emma Woodhouse with Mr. Knightley. Marianne Dashwood and Willoughby dance together in an uncontrolled and improper manner during the first couple of months of their acquaintance, which, retrospectively, anticipates the unfavourable outcome of their relationship. On other occasions, when, for instance Marianne and Anne Elliot are mourning their relationships with, respectively, Willoughby and Wentworth, they are not psychologically able to dance.

In this section, then, I will look at the emotional implications and the physical aspect of dancing, as well as the spatiotemporal specificity of the ballroom as a living space. In addition, I will analyse the place of dance within different cyclical structures of human life and its impact on linear events.

Dancing Space-Time

A ball can be described as a particular spatiotemporal unit inserted in an everyday living space that is emotionally as well as physically inhabited by the participants. Unlike walking, dancing requires a specifically transformed interior space. As preparation for the Westons' ball in *E* illustrates (*E* 267–268) this is especially problematic for private balls when an upper-middle class house has to accommodate a set of dancing couples. Some houses, however, seem to be more "flexible" than others. Spontaneous balls are organized at Lucas Lodge in *P&P*, at the Coles' house in *E*, at Uppercross in *P*, and at Mansfield Park in *MP*, while Sir Thomas is away. The first three houses are inhabited by sociable people who have only recently acquired genteel status. Their home-spaces do not carry the weight of preceding generations, thus "one end of the room" (*P&P* 27) is easily turned into an improvised ballroom for the youngest Miss Bennets and Miss Lucases with three eager officers, and there is a "rapid [...] clearing away" (*E* 248) to provide some dancing space for the Coles' dinner party guests. In Uppercross Great House "an unpremeditated little ball" is not an unusual event (*P* 51), which reflects the Musgrove family's spontaneous and friendly character: they are said to be "visited by every body, and [have] more dinner parties, and more callers, more visitors by invitation and by chance, than any other family" (*P* 50–51). On the other hand, the improvised dance in Mansfield Park happens significantly during the absence of Sir Thomas and the stay of Mr. Bertram with his friend, who plan to "violate" the traditional trans-generational house space by transforming it into a theatre. The main reason for this ball "thought only of [*sic*] the afternoon" (*MP* 137) is the sudden possibility to have enough male partners to form five couples, which would give Mary and Henry Crawford some semi-proper opportunities to flirt respectively with Edmund and Maria. Overall, the spontaneous balls seem to be inspired by the desire for physical proximity springing from "wilder" and less prudent characters, such as Lydia Bennet, Frank Churchill, Louisa Musgrove, and the Crawfords.

During these improvised balls in private houses the everyday space undergoes more emotional than physical transformation. The room itself is minimally altered and the principal goal is the pleasure of dancing, especially with a sexually attractive partner. However, not every attempt to organize a spontaneous dance is successful. Mr. Darcy, contrary to Frank Churchill or Henry Crawford, fails to convince the woman he likes that under the cover of irony he is sincerely desirous to "seize such an opportunity of dancing a reel" (*P&P* 56). The prejudice that Elizabeth has regarding Mr. Darcy, makes her look for provocative motives in his invitation and adopt a defensive attitude.[9] Thus, she fails to realize that Mr. Darcy's offer might have risen from the wish to share a closer physical experience with her.

A private ball organized in advance is a very different affair from the improvised ball. The ballroom and all that surrounds it, like the room for the supper or the card-room, are prepared for the occasion. It is a ritualized event with rules and codes that require a formal space. The private balls are organized by the wealthier families in the novels: in *S&S* the Autumn balls are given by Sir John, in *P&P* there is a ball at Netherfield Park, and in *MP* a ball is organized for Fanny to mark her coming out. We have a special case in *E*, where the Westons, under pressure from Frank, organize a private ball in a rented space and not in their house, which is considered too small for dancing. Evidently, holding a proper private ball is expensive, and many families cannot afford it. For example, we never hear anything about a ball organized at Longbourn, though Mrs. Bennet would surely have been interested in hosting such an event. However, some of the wealthiest and the oldest families also never hold a dancing event. There is no ball at Northanger Abbey, Pemberley, Donwell Abbey, or Hartfield. There are dinners, luncheons, and tea parties, but we do not see anybody dancing in these spaces. Inclination for dancing seems to require a certain lightness of spirit and inner spontaneity that serious Mr. Darcy, Mr. Knightley, and General Tilney, as well as anxious Mr. Woodhouse do not have.

The third category of balls in the novels are the public balls in Bath and the assemblies in Meryton (*P&P*) and D. town in Surry (*W*). The ball in London in *S&S*, though private, is so big, that it could be compared to a public ball. These spaces are represented as crowded, strange, even hostile, especially, when a heroine has very few acquaintances, as is the case for Catherine Morland, the Miss Dashwoods and Emma Watson. Catherine Morland's first ball in Bath is the most disappointing and is comparable to tragicomic swimming against the flow without ever reaching the promised dancefloor:

[9] According to D. A. Miller, Elizabeth is wrong when she supposes Darcy has "a very satirical eye" (*P&P* 24) for he is "far too stiff for that" (43). Miller claims that: "Elizabeth is projectively mistaking him for the only person she knows who does have such an eye, namely, her own father, who is continually casting it, to disdainful effect, on the female members of the household" (43). If that is the case, Darcy must be incapable of making an ironic offer. However, we can clearly hear irony in some of his other comments, for instance, his remark on Wickham during the two dances he shares with Elizabeth (102–103).

> They [Catherine and Mrs. Allen] were not long able [...] to enjoy the repose of the eminence they had so laboriously gained.—Every body was shortly in motion for tea, and they must squeeze out like the rest. Catherine began to feel something of disappointment—she was tired of being continually pressed against by people, the generality of whose faces possessed nothing to interest, and with all of whom she was so wholly unacquainted, that she could not relieve the irksomeness of imprisonment by the exchange of a syllable with any of her fellow captives. (*NA* 14)

Catherine, who is used to open spaces of the countryside, here has a claustrophobic experience "squeezed," "pressed," and "imprisoned," by an anonymous crowd. She is not noticed, let alone distinguished by anybody, and is not able to individualize the faces that surround her. This is only the first of many instances of the narrative where her expectations clash with the disappointing reality. The passage also highlights the importance of personal social bonds in order to really integrate the codified space of a public ballroom. In complete contrast to Catherine Morland, the Miss Bennets in *P&P* know the families of Meryton too well, and only the new arrivals—Mr. Bingley, his sisters, and his friend—add some real interest to the ball.

On another level, the ballroom space is divided into two areas: the space for dancing, or active space, and the space for watching people dance, or the passive space. According to ballroom etiquette, "a gentleman, single or married, was expected to invite ladies who were available to dance" (Fullerton, 85). However, there are several gentlemen in the novels—Captain Tilney, Mr. Darcy, Mr. Bertram, Mr. Knightley, Mr. Elton—who break this rule and prefer to stand in the passive space on the outskirts of the dance floor.[10] Fullerton pragmatically notes that men had better opportunities of "ogling" dancing women if they stood here at the side, and that women, being aware of that, tried to look their "absolute best" (24). She adds that for this reason "Mr Darcy does not have to make himself available as a dance partner (or as a husband, which is really what he is signalling when he refuses to dance), but can instead stand back and, as a voyeur, inspect the available talent until tempted by some Georgian beauty simply too appealing to resist" (23–24). Mr. Darcy, however, contradicts Fullerton's opinion later in the novel, when he declares to Elizabeth and

[10] For D. Siegel and R. Handler, who comment on the etiquette at the time, men had a moral obligation to choose a dancing partner "among the possible choices" (326), and they assert that at balls "each woman may be treated differently by each man, but women must be treated alike by men" (326).

his cousin Colonel Fitzwilliam that he is too shy to dance with strangers (*P&P* 196). Mr. Knightley is also definitely not a voyeur. Nevertheless, he is looking from the side at dancing Emma (*E* 353), already aware of his feelings for her (as the reader will find out retrospectively) and he is attentive to the general situation in the ballroom, for he notices when Harriet is snubbed by Mr. Elton. On the other hand, Lord Osborne, in *W*, seems to correspond to Fullerton's voyeur. First, he sends his friend Tom Musgrave to invite Emma to dance in order to get a good opportunity to observe her closely (*W* 101). When his first plan does not succeed, for Emma is asked to dance by Mr. Howard, he positions himself "at Howards elbow during the two dances" (*W* 103).

In contrast to the watching men, the women who are seated during a dance, do not normally choose this situation. In one of her letters Jane Austen mentions having danced nine dances out of twelve and having been "merely prevented from dancing the rest by the want of a partner" (60). Similarly, her heroine Elizabeth Bennet is "obliged, by the scarcity of gentlemen,[11] to sit down for two dances" (*P&P* 11), Catherine Morland sits during her first ball in Bath because she does not know anybody (*NA* 16), and, later, because her partner John Thorpe is late for the dance (*NA* 48), Fanny Price is neglected by her cousin Tom Bertram (*MP* 139), Harriet Smith is rejected by Mr. Elton (*E* 354), and Anne Elliot, stuck in the role of spinster, plays for the dancers while her "eyes fill with tears" (*P* 77).

The surrounding passive space of the dance floor is, thus, an emotionally intense space. Those who dwell on the edges either hope to be in the middle or wish to get closer to somebody who is dancing there. In a way, the active dancing space is a dream space, comparable to the concept of a dream house that I discussed in the previous chapter, or, more precisely, it represents a space which is also a path leading to the construction of the dream home through the successful choice of the life-partner.

In contrast, the active dance space can also be felt as a temporary trap (or nightmare): once in the dance, there is no way out until the end of the traditional two dances, which, given Henry Tilney's comparison, could also be seen as an emblem of a bad marriage. Thus, Catherine Morland, compelled to refuse Henry Tilney's invitation and to dance

[11] Fullerton explains that lack of dancing partners came from the "real shortage of men" (95) because of the war with France: approximately 100,000 men were in the army, and 130,000 in the navy. Many of them got killed, others spent their time fighting in Europe or at sea. Moreover, the expansion of the British Empire "was also starting to lure young men away" (96).

with Mr. Thorpe, feels "separated from all her party, and away from all her acquaintance" (*NA* 50) and can only helplessly continue to look "very often towards the part of the room where she had left Mr Tilney" (*NA* 50). Similarly, Elizabeth Bennet is stuck with Mr. Collins, who had reserved the first two dances with her in advance, and she experiences "all the shame and misery which a disagreeable partner for a couple of dances can give" (*P&P* 101). Finally, in *MP*, Fanny Price is "a great deal too much frightened to have any enjoyment" (*MP* 321) when she opens the ball with Mr. Crawford under the observing eyes of all the other guests.

The active dance space also combines the phenomena of cyclicity and linearity. The strict choreographic pattern previously learnt and then reproduced by the dancers across the floor, repetitive steps, hops, or turns, and the circular rotation of couples dancing their way down the line creates an image of cyclicity par excellence.[12] The traditional succession of dances during the ball and, in general, the regular chain of public balls together with seasonal series of private ones reflect the social cyclicity of genteel living. Even the conversation between the dancing partners could follow a pre-determined model, as Elizabeth Bennet and Henry Tilney mockingly demonstrate in *P&P* (102) and *NA* (18) respectively. However, the emotional state that accompanies this physical and social movement may lead to some important linear events and decisions.

The relationship between dancing and time is also multi-layered. First, there is the question of seasonal cyclicity. As I pointed out in my first chapter, the country balls are organized in Autumn during the shooting season and may last until Christmas. In Austen's unfinished novel *The Watsons*, we even get the precise starting date: "the first winter assembly in the town of D. in Surry was to be held on Tuesday October the 13th" (*W* 79). In *S&S* we have balls in October organized by Sir John, in *P&P* the Meryton assembly is in mid-October, while the Netherfield ball is on 26 November, and there are also autumnal dances at Mansfield Park and Uppercross. Thus, these balls often correspond to the time of new beginnings in the lives of the heroines as I discussed in my first chapter. The balls in London and Bath take place during the Season from about February to May.[13] Thus we find Catherine Morland dancing in the Upper and Lower Rooms in February, and meeting Henry Tilney there. On the other hand, the only heroines who have a chance to participate in the

[12] See Fullerton (99–112) and Wilson (26–34) for details concerning different country dance patterns.

[13] See Davidoff, 21.

London season—Elinor and Marianne Dashwood—are not in the right emotional state for dancing. Young Fanny Price, however, indirectly experiences the season of balls through the conversations with her aunt: "as to her cousins' gaieties, she loved to hear an account of them, especially of the balls, and whom Edmund had danced with" (*MP* 40). The only exception to the seasonal rule of balls is Highbury in *E*. The improvised dance at the Coles' is in February, while the Westons' ball that was first talked of in Winter takes place only in May.

At another level, dancing time relates to a particular stage of human life. Most of the dancers in the novels are young and single. A married person dancing has a disinterested reason: in *E*, Mr. Weston has a moral obligation to open the ball with the new bride Mrs. Elton, while Mrs. Grant joins the improvised dance in *MP* in order to complete the fifth couple.[14] Bearing in mind that the main goal of dancing was to meet a marriageable partner, its quasi-exclusivity for the young seems logical. However, the physical and emotional aspects might also be important. In one of Austen's later letters we can witness her significant change of attitude towards balls. In a letter written on 14 October 1813 she rejoices at not having gone to a ball: "I was very glad to be spared the trouble of dressing & going & being weary before it was half over" (237). Though not married, Austen at 38 must have been considered as too old for active dancing and, even more so, for meeting a man to marry. This might be one of the reasons why a ball has lost its appeal. Moreover, the letter indicates that she is feeling less physically strong than she was in her twenties.

On the other hand, young married men were expected to dance if there were any single ladies left without partners (Fullerton, 85), which is one of the reasons twenty-seven year old Mr. Elton's declaration of being "an old married man" (*E* 354) and his refusal to dance with Harriet Smith are so offensive.

At this point I would like to redirect my focus to the space/time of the ballroom, which is always very densely packed in physical and psychological terms. The dancing partners advance rapidly in their mutual understanding or misunderstanding and form their opinions about each other through this exclusive occasion for a private conversation. At the same time, they are completing "the demands of the dance" (*NA* 77) and, occasionally, observe other couples or the people who are watching them.

[14] D. A. Siegal and R. Handler point out that married people's "dance activities serve frequently to facilitate the dancing of those who should dance, that is, the young and marriageable" (326).

However, as they cannot easily leave the dance space before the end of the set, the active ballroom space resembles a closed space/time cell that functions according to its internal rules of pattern and rhythm. Those who watch, on the other hand, stay in the objective outside space/time, where they admire, criticize, and speculate about possible attachments between the dancers. Thus, a ball is comparable to a nucleus, a knot of the forthcoming linear events that will expand and evolve afterwards. In *NA*, *P&P*, *MP*, and *E* dancing allows for the propelling forward of the action, as well as the clarification of the relationships between its participants. In contrast, in *S&S* and *P*, dancing parties or balls highlight Marianne's and Anne's physical and emotional inaccessibility: Marianne is exclusively concentrating on Willoughby, while Anne is endorsing the role of the musician.

The Joy of Dancing and Other Emotions

Except for the earlier-mentioned instances of especially unpleasant partners, a ball is usually referred to in the novels as a joyful experience that is always looked forward to. Dancing means following a repetitive rhythm and pattern, which inevitably accelerates the heart-beat and the breath, thus making the dancers more conscious of their moving bodies, while the music must naturally appeal to some primary source of instinctive happiness—the joy of moving and feeling alive. This is suggested when the Miss Musgroves in *P* are said to be "wild for dancing" (*P* 51), Emma Woodhouse "leads off the dance with genuine spirit and enjoyment" (*E* 248), Maria Bertram's eyes are "sparkling with pleasure" during the improvised ball at Mansfield Park (*MP* 138) and Miss Watson's "delight in a ball [is] not lessened by a ten years enjoyment" (*W* 79). Harriet Smith, when rescued by Mr. Knightley, unites emotional and physical excitement in her dancing: "she [bounds] higher than ever, [flies] farther down the middle, and [is] in a continual course of smiles" (*E* 355). Jane Austen expresses similar positive emotions in her letters when she writes of her dancing experiences: she dances with "much satisfaction," would be ready to "dance for a week together" (30) and is "full of joy" knowing that her sister has had an opportunity to dance and, thus, to enjoy herself (30). The lively Elizabeth Bennet, on the other hand, though anticipating with pleasure the possibility of dancing (*P&P* 97), is never given the opportunity to dance without prejudice, resentment, or mortification during the time of the narrative. Austen heroines thus go through various emotions during the time of a ball.

In general, what differentiates dancing from walking, is the continuous tension between the pleasure of dancing and the frustration at not having the right partner (and, sometimes, any partner at all) or enough time to enjoy the dance with the right man. Walking most often appeases agitated emotions and brings a certain degree of serenity, whereas dancing, on the contrary, excites the emotions and gives plenty of new events or observations that will have to be thought over later, possibly, during the walks. Moreover, walking is a common daily activity, while a ball, for Austen's heroines, is mostly a special occurrence dependent on the time of year and the constraints of organization. These young women consequently have high expectations for every ball that they have the opportunity to attend. As I have mentioned before, dancing rhymes with the possibility of meeting a life-partner and the young potential brides feel the pressure to make the most of it.

NA offers us the most scenes of dancing with five public balls in Bath which Catherine Morland attends. The very first ball, however, is a frustrating experience for Catherine, because, as she does not know anybody there she cannot dance (*NA* 14–15). Her second ball is much more successful for the master of ceremonies introduces Henry Tilney who asks her to dance (*NA* 17). From this moment on Catherine does not only wish to dance, she wishes to dance with him, and any other partner, especially anyone as egocentric, irksome and arrogant as John Thorpe, annoys her, as I have already mentioned. During her fourth ball she hides and avoids Mr. Thorpe as much as possible in order to keep herself disengaged for Henry Tilney, and "though she could not, dared not expect that Mr Tilney should ask her a third time to dance, her wishes, hopes and plans all centered in nothing less" (*NA* 72). Thus, the main emotions that Catherine experiences in the ballroom are agitation, anxiety, and fervent anticipation. Actually, she is orienting all her future there, by trying to put herself in the way of the man she prefers and avoid the unsuitable candidate, and she does it with great energy and determination. The only completely blissful and relaxed moments are the actual dances with Henry Tilney, which, according to the etiquette could not be more than two during one ball: "Catherine, meanwhile, undisturbed [...] of any evil at all, except that of having but a short set to dance down, enjoyed her usual happiness with Henry Tilney, listening with sparkling eyes to everything he said" (*NA* 134). Thus, during the dances with Tilney she is dwelling in a subjective space-time bubble and is detached from the surrounding world. The time she can spend close to him seems far too short, as is the case for Emma Watson in *W* after her two enjoyable dances with Mr. Howard (*W* 103).

In *S&S* dancing is accorded very little place. And although we are told that parties are regularly organized by Sir John, there is no detailed description of them. Nevertheless, we learn that "this was the season of happiness to Marianne" (*S&S* 64), and we are given the opportunity to measure the extent of Marianne's passion for Willoughby:

> When he was present she had no eyes for any one else. [...] If dancing formed the amusement of the night, they were partners for half the time; and when obliged to separate for a couple of dances, were careful to stand together and scarcely spoke a word to any body else. (*S&S* 64)

The passage shows that dancing is for Marianne and Willoughby mainly an occasion to be close together, and they defy the etiquette of the ballroom that demanded the changing of partners after two dances.[15] In fact, they invert the rule by separating for two dances and dancing the rest of the time together. Moreover, they neglect the customary decorum by not speaking to anyone else. These Autumn parties constitute a contrast to the ball Marianne and Elinor go to in London. In the Autumn parties, Marianne happily dances all the time with the same partner, while in London she does not dance at all and is practically disowned by Willoughby. This also illustrates Marianne's extremely large emotional range: she passes from perfect happiness and overt demonstration of her love to inconsolable despair. In contrast, it is never mentioned in the novel if Elinor dances. We might presume that she does so during the Autumn parties, for she is young and marriageable; however, as Edward, with whom she is in love, is away, dancing might have little "strategic" interest for her and, thus, for the narrative.

All three balls of *P&P* are concentrated in the first volume. For Jane Bennet, this is a happy time, when she begins to construct her relationship with Mr. Bingley, which is generally expected to lead to a marriage proposal. And even though there will be several months delay before the proposal actually comes, the definitive inner transformation takes place in the period prior to and during the Netherfield ball. Similarly to Marianne Dashwood and Willoughby, though less extremely, Jane and Mr. Bingley spend time together dancing and end up "detached from the rest" talking "only to each other" (*P&P* 115) while they are waiting for the carriage to take Jane and her family home. Elizabeth sees "true affection" (*P&P* 110) between them and imagines their forthcoming happy marriage.

[15] Fullerton explains that as dancing was a social activity, a couple dancing always together would neglect "their duties as members of that community" (84).

For Elizabeth, who, as I have discussed, enjoys physical exercise, one would expect dancing to be an especially pleasant experience. However, during the time of the narrative she does not have luck in it. Though in Meryton she is said only to sit through two dances (*P&P* 11), Mr. Darcy's remark that she accidentally[16] hears "mortifies" her pride (*P&P* 21) and lays the foundation for her prejudice towards him. During the spontaneous dance at Lucas Lodge she refuses Mr. Darcy's invitation in order to appease that mortified pride. Finally, during the Netherfield ball, she feels frustrated and disappointed, because her expectations of dancing with Mr. Wickham and conquering "all that remained unsubdued of his heart" (*P&P* 100) are shattered by his absence. Instead, she is forced to partner Mr. Collins and Mr. Darcy or not dance at all. Like Catherine Morland and the intrusive John Thorpe, Elizabeth has to bear the presence of the tenacious Mr. Collins and "the moment of her release from him [is] exstacy" (*P&P* 101). However, unlike Catherine, who immediately recognizes in Henry Tilney her ideal future husband, Elizabeth is far from being able to picture herself as Mr. Darcy's bride, thus, dancing with him equals a punishment. Their bouncy conversation, with several silent gaps when they go "down the dance" (*P&P* 103, 105) and with erotic tension which might be spotted by an attentive first-time, or an experienced, reader, does not give her pleasure and does not alter her negative opinion about him. However, on Mr. Darcy's side an important linear transformation, which began a few weeks earlier, continues, and by this time in his "breast there was a tolerably powerful feeling towards her" (*P&P* 105).

There are two dancing occasions in *MP*, and both bring their spell of happiness to the participants: Miss Bertram's eyes are "sparkling with pleasure" (*MP* 138), Miss Crawford is in "gay spirits" (*MP* 324), and Fanny has "a good deal of enjoyment in the course of the evening" (*MP* 323). However, Mr. Crawford's tenacious attention is unpleasant to her (*MP* 323) and Fanny's happiness springs from the expectation and, later, the opportunities to dance, talk, and, perhaps most surprisingly, "have the luxury of silence" with Edmund and from "being the friend" (*MP* 323) that he could rely on:

> She was happy in knowing herself admired, and she was happy in having the two dances with Edmund still to look forward to, during the greatest part of the evening, her hand being so eagerly sought after, that her indefinite

[16] Mr Darcy, however, might have meant for her to hear it for he turns around and looks at Elizabeth "till catching her eye" (12).

> engagement with *him* was in continual perspective. She was happy even when they did take place; but not from any flow of spirits on his side, or any such expressions of tender gallantry as had blessed the morning. His mind was fagged, and her happiness sprung from being the friend with whom it could find repose. (*MP* 323)

The passage gives an account of various forms and shades of Fanny's happiness, pointing at the manifold nature of it. Symbolically, this passage also predicts Fanny's and Edmund's "indefinite engagement" and their pathway towards a common future. At the moment it is still "in perspective"; however, Fanny's constancy of feeling and steadiness of character will create a "welcoming harbour" for Edmund after his emotionally tiring and disappointing relationship with Mary Crawford.

Unlike the other heroines, the main vexation that Emma Woodhouse has to endure in a ballroom is not being the one to open the ball at the Crown: Mrs. Elton, as the new bride, gets this honour, which is "almost enough to make her [Emma] think of marrying" (*E* 352). Nevertheless, as balls are especially rare in Highbury "Emma was smiling with enjoyment, delighted [...] to feel that she had so many hours of unusual festivity before her" (*E* 352). Noticeably, her attention during the dance is regularly turned towards Mr. Knightley. She looks at him standing between the older men, which must emphasize his sexual attraction, admires his figure as well as his manners, and is regularly trying to catch his eye (*E* 352–353). The moment she sees him asking Harriet, slighted by Mr. Elton, to dance, is especially intense:

> Never had she been more surprised, seldom more delighted, than at that instant. She was all pleasure and gratitude, both for Harriet and herself, and longed to be thanking him; and though too distant for speech, her countenance said much, as soon as she could catch his eye again. (*E* 355)

The positive and strong emotions that Emma experiences convey the depth of her feelings for Mr. Knightley of which she is still not aware. She has a chance to communicate her appreciation of his behaviour through the distance that separates them because Mr. Knightley is constantly, and more consciously, watching her too. This scene emotionally prepares their actual dancing together that comes slightly later in the narrative (*E* 358). Although the reader does not get the opportunity to see them partnering each other, this moment marks an important linear transformation in their emotional history: Emma and Mr. Knightley for the first time are sharing space/time as a potential couple and not only as old friends. We get a similar picture in *MP*, when Fanny dances with Edmund, except that Edmund, unlike Mr. Knightley, is not yet capable of seeing his cousin as a potential lover.

Of all Austen's heroines Anne Elliot is the only one who does not dance during the time of the novel, and not because she has not got the opportunity, but by choice: "she has quite given up dancing" (*P* 78). Segal and Handler remark that her "sad slide into spinsterhood is signalled by the fact that her friends have come to depend on her 'services' as a musician for their dances" (*P* 326). Thus, as Anne has given up on love, she has stopped dancing. Nevertheless, as I mentioned earlier, it is an emotionally trying moment for her, to be playing for the amusement of Captain Wentworth with the young and pretty Miss Musgroves. According to Wilson, however, Anne has her indirect "dancing" moment at the very end of the novel when she gets engaged to Captain Wentworth. She comments that the "final scene of their reconciliation employs the language and imagery of dance in a way that encourages readers to see Anne and Wentworth walking side by side as an event equally momentous and sexually charged as the dancing of Elizabeth Bennet and Mr Darcy" (4–5). Hence, "dancing" does not only happen in the ballroom, "dancing spirits" (*P* 261) that Anne experiences at this moment, as Catherine Morland does on her way home after having danced with Henry Tilney (*NA* 79), metaphorically represent the experience of such happiness as dancing procures.

To conclude, walking and dancing, in the novels, generally evolve in emotionally paved space/time. Walking is most often initiated by an emotional storm and need for quiet rumination in an outside space and inner time, while dancing generates strong emotions and accelerates events in a strictly determined, ritualized space and time. However, both acts mark and/or initiate important linear events in the everyday cyclicity of human life.

The novels where we find the most numerous instances of walking are *S&S*, *P&P*, and *E*; dancing is more present in *NA* and *P&P*. In *MP* and *E* there are only two dancing occasions, however, both of them are important for the development of the plots and have linear quality. Hence, *P* features the only heroine who is only slightly or not at all engaged in the physical activities of walking and dancing. Moreover, the novel is full of images of illness, tiredness, the failing body; even the young characters—Louisa and little Charles—suffer physically traumatizing accidents followed by slow recoveries. However, during the narrative Anne accomplishes an inner journey of freeing herself from the cyclical mourning and reconstructing her future, which restores her "bloom and freshness of youth" (*P* 112). Thus, the last novel of Austen seems to shift focus towards psychological as well as physical transformations accomplished by mental force.

Works Cited

Primary Sources

Austen, Jane. *Northanger Abbey*. Ed. Barbara M. Benedict and Deidre Le Faye. Cambridge: Cambridge University Press, 2006a.

Austen, Jane. *Sense and Sensibility*. Ed. Edward Copeland. Cambridge: Cambridge University Press, 2006b.

Austen, Jane. *Pride and Prejudice*. Ed. Pat Rogers. Cambridge: Cambridge University Press, 2006c.

Austen, Jane. *Mansfield Park*. Ed. John Wiltshire. Cambridge: Cambridge University Press, 2005a.

Austen, Jane. *Emma*. Ed. Richard Cronin and Dorothy McMillan. Cambridge: Cambridge University Press, 2005b.

Austen, Jane. *Persuasion*. Ed. Janet Todd and Antje Blank. Cambridge: Cambridge University Press, 2006.

Austen, Jane. *Later Manuscripts*. Ed. Janet Todd and Linda Bree. Cambridge: Cambridge University Press, 2008.

Austen, Jane. *Letters*. Ed. Deidre Le Faye. 3rd edition. Oxford, New York: Oxford University Press, 1995.

Secondary Sources

Amato, Joseph A. *On Foot: A History of Walking*. New York: New York University Press, 2004.

Davidoff, Leonore. *The Best Circles. Society, Etiquette and The Season*. London: Croom Helm, 1973.

Fullerton, Susannah. *A Dance with Jane Austen*. London: Frances Lincoln Limited, 2012.

Hanna, Judith Lynne. *To Dance is Human. A Theory of Nonverbal Communication*. Chicago: University of Chicago Press, 1979.

Johnson, Claudia L. *Jane Austen's Cults and Cultures*. Chicago: University of Chicago Press, 2014.

Jones, Hazel. *Jane Austen & Marriage*. London: Continuum, 2009.

Le Faye, Deirdre. *Jane Austen. The World of her Novels.* London: Frances Lincoln, 2002.

Lefebvre, Henri. *The Production of Space.* Trans. Donald Nicholson-Smith. Oxford: Blackwell, 1991.

Miller, D. A. *Jane Austen or the Secret of Style.* Princeton: Princeton University Press, 2003.

Moretti, Franco. *Atlas of the European Novel 1800–1900.* London: Verso, 1998.

Pidduck, Julianne "Of windows and country walks: frames of space and movement in 1990s Austen adaptations." *Screen* 39.4 (1998): 381–400.

Pool, Daniel. *What Jane Austen Ate and Charles Dickens Knew.* New York: Touchstone, 1993.

Segal, Daniel A. and Richard Handler. "Creative Dance and Dramatic Sensibility in Jane Austen, Ethnographer" in *Man, New Series* 24. 2 (1989): 322–339.

Sulloway, Alison G. *Jane Austen and the Province of Womanhood.* Philadelphia: University of Pennsylvania Press, 1989.

Tanner, Tony. *Jane Austen.* London: Macmillan, 1986.

Wallace, Anne D. *Walking, Literature, and English Culture.* Oxford: Clarendon Press, 1993.

Wilson, Cheryl A. *Literature and Dance in Nineteenth-Century Britain.* Cambridge: Cambridge University Press, 2009.

Worsley, Lucy. *Jane Austen at Home.* London: Hodder & Stoughton, 2017.

CHAPTER 5

Conclusion

In this study I have shown the complexity of the spatiotemporal architectures in Austen's novels. The three chapters look at different ways the novels structure space/time in relation to the protagonists and their psychological development.

To begin, I pointed out how natural cycles of the changing seasons and of human life merge with cultural cycles of the liturgical and agricultural calendar, social seasons, and everyday rituals. Linear events emerge at pivotal points during the year and move characters from one life stage onto another making them start a new cycle. Autumn with its arrivals and departures, its losses and newborn hopes represents the new beginnings, mirroring at once the decaying of nature and the sowing of seeds for the next Spring. Winter is recurrently the time for physical and psychological gestation and reflection, Spring brings news and hope, while Summer is the month of final revelations and denouements. The year has pivotal points in Christmas, Easter, and Mid-Summer, which are more or less important in different novels, and which unsettle and transform the spatiotemporal constellation of the narratives. As I pointed out too, the seasonal structure of the six novels evolves from a simple binomial Winter-Spring structure in *NA* to a structure of four seasons unequally divided, in *S&S* and *P&P*, and an almost equally divided structure in *MP* and *E*. In her last novel, Austen fits her plot into an incomplete year again, as in *NA*, but she adds depth to its temporal structure with the omnipresent line of the heroine's past.

R. Baublytė Kaufmann, *The Architecture of Space-Time in the Novels of Jane Austen*, https://doi.org/10.1007/978-3-319-90011-7_5

During the time of the narrative, the heroines accomplish their psychological and physical journeys through the four types of life spaces connected to the stages of their psychological development: childhood, temporary, dream, and adult homes. All the novels convey the idea that home is represented by affective bonds more than by a particular house. However, in *P* Austen goes further, leaving her heroine with no fixed home of her own, but with a man she loves and is ready to follow wherever he goes. This suggests a progressive idea of women freed from the binding relationship with a fixed place and the attendant restricted physical, social, and emotional mobility.

Spatiotemporal patterns and rhythms are also expressed through the physical and psychological movement that is walking and dancing. Both acts are omnipresent in the novels, though not equally. The most prominent Austenian walkers are: Marianne Dashwood, who is mainly an emotional walker; Elizabeth Bennet, who gets physical pleasure and emotional relief during her walks; and Emma Woodhouse, who is a perfect example of a social walker. As well as a physical exercise and means for privacy, walking offers space/time for solitary reflection and the calming of agitated emotions. Dancing, on the other hand, often generates strong emotions ranging from joy and happiness to frustration and disappointment. Moreover, balls are limited to a very specific space/time, as much in terms of seasons as in terms of years, thus they represent a nucleus of the potential linear events that could lead the heroines towards their future happiness.

In each of my chapters I have separated and analysed the recurring spatiotemporal patterns. It is important to remember, however, that in the novels these patterns are superposed in a multi-layered structure. There are certain moments in the narratives that are thus particularly charged with meaning, such as Marianne Dashwood's autumnal farewell to Norland and her nearly fatal illness the following Spring (*S&S*) discussed in Chaps. 2, 3, and 4; Elizabeth Bennet's solitary moments during her Spring visit to Kent (*P&P*) discussed in Chaps. 2, 3, and 4; and the walk to Winthrop in Autumn (*P*) discussed in Chaps. 2 and 4. The different spatiotemporal patterns discussed in separate chapters thus work together to charge the moments with intensity of meaning.

Finally, I have shown that Jane Austen's novels are constructed on deep internalized cycles. The changing seasons anchor human subjects in their geographical space and connect them to their ancestors, influencing their everyday actions, their thoughts and their dreams. Their life spaces reflect their time of life and changes brought about by inner transformations, while movement, such as walking and dancing, builds connections between their physical, social, and emotional life, hence enabling them to negotiate the inevitable linearity.

Even though today the seasons appear to be losing their distinctive characteristics and the fundamental, reassuring role they used to have, "our daily experience of the climate and the calendar still bears the deep stamp of centuries of culture, history, religion, politics, and agriculture" (Groom 2014, 22). The cyclicity of seasons coexisting with sudden or long-expected linear events, the life spaces transformed through time, or the rhythms of moving bodies and minds are still present in collective as well as individual consciousness. With a tinge of nostalgia we—at least readers in the Western world—sense these fundamental structures in Austen's novels and it makes us feel "at home"—in some emotionally secure space between original and dream home—when we read them.

Work Cited

Groom, Nick. *The Seasons: An Elegy for the Passing Year.* London: Atlantic Books, 2014.

Index[1]

A
Austen, Jane
letters, 9, 138, 157, 161
novels, 1–5, 9, 11–14, 19–22, 20n3, 27, 35, 48, 75, 76, 78, 81, 84, 89, 104, 107, 118, 122, 133–135, 156, 169, 170, 173–175
plots, 9, 19, 21, 25, 29n7, 42, 76, 78, 173
rereading, 1

B
Ball
in Austen's letters, 9, 11, 163
emotions, 13, 42, 75, 93, 156, 158, 163–170, 174
female/male, 43, 155
private, 38, 93, 158, 159
public, 5, 58, 127, 156, 159, 162, 165
Bath, 4, 20, 26, 27, 33, 34, 36, 52, 61, 70, 72–74, 76, 78, 82, 86, 87, 93, 116, 117, 127, 128, 150–153, 156, 159, 161, 162, 165

C
Change, 6, 7, 22, 22n5, 23, 26, 34, 36, 37, 41–43, 47, 48, 50, 52–57, 60, 62–65, 67–72, 76–78, 82, 84, 93, 101, 113, 133–170
Christmas, 4, 20, 20n3, 25, 26, 28, 32, 33, 38, 43, 44, 46, 49–51, 58–61, 64, 66, 72, 77, 78, 125, 162, 173
City, 39, 108, 150
Country, 13, 20, 24, 27–29, 34, 40, 47, 51, 71, 75, 91, 94, 108, 109, 119n8, 135, 144, 151, 162, 162n12

[1]Note: Page numbers followed by 'n' refer to notes

R. Baublytė Kaufmann, *The Architecture of Space-Time in the Novels of Jane Austen*, https://doi.org/10.1007/978-3-319-90011-7

Cycles, 4–7, 9, 19–78, 84, 89, 92, 96, 98, 104, 111, 115, 119, 120, 145, 173
Cyclicity, 3–5, 3n2, 7, 9, 14, 21–23, 25, 26, 28, 29, 33, 36–39, 41, 43, 46–49, 51, 52, 54–57, 62, 64–66, 68, 71, 72, 74, 84, 89, 95, 99, 104, 108, 119, 128, 133, 138, 140, 147, 149, 152, 162, 169, 175

D
Dancing
  emotions, 8, 13, 134, 156, 157, 161–170, 174, 175
  space, 8, 12, 133, 134, 157–165, 169, 174, 175
Doors
  closed, 125
  opened, 123

E
Easter, 4, 20, 20n3, 27, 28, 38, 40, 45, 53, 55, 173
Emma, 12, 13, 57–70, 110–115
Emotion
  agitated, 8, 35, 134, 145, 154, 165, 174
  calm, 41, 64, 134, 142, 143, 147
  contradictory, 8, 42
  transformation, 2, 37, 41, 54–56, 70, 84, 95, 96, 101, 158, 168, 174, 175

H
Homes
  adult, 7, 82, 84, 88, 89, 95, 99, 104, 105, 107, 108, 110–112, 114–118, 149, 174
  dream, 7, 10, 34, 82–84, 87–89, 94, 95, 102, 103, 107, 109, 110, 113, 114, 117, 118, 161, 175
  original/childhood, 4, 5, 7, 10, 25, 28, 50, 54, 71, 82–86, 89, 91, 92, 94–96, 98, 99, 105, 107, 108, 110–112, 114–119, 127, 142, 149, 152n6, 154, 174, 175
  temporary, 26, 36, 37, 70, 71, 81–83, 86, 92, 94, 96, 100, 101, 107, 108, 110, 112, 113, 115, 116, 174

L
Linear events, 8, 10, 19, 22, 25, 29, 32, 41, 42, 52, 54, 56, 58, 61, 62, 74–76, 82, 154, 157, 162, 164, 169, 173–175
Linearity, 2, 4, 7–9, 14, 23, 34, 37, 39, 42, 48, 53, 63, 65, 66, 68, 72, 75, 133, 137, 140, 147, 149, 152, 162, 175
London, 4, 20, 26–29, 33, 34, 38–41, 45, 51–53, 55, 58, 61, 76, 77, 82, 83, 91–93, 101, 108, 110, 128, 144, 146, 159, 162, 163, 166
Lyme Regis

M
Mansfield Park, 9, 11–13, 50–57, 105–110
Michaelmas, 4, 19, 20, 20n3, 42, 49, 70
Mid-summer, 173
Movement, 3, 5, 6, 8, 13, 14, 22n5, 23, 24, 27, 28, 36, 73, 77, 133–170, 174, 175

N

Northanger Abbey, 4, 11, 21, 32–36, 85–89, 149

P

Patterns, 2, 4, 5, 10, 14, 21, 29n7, 76, 78, 152, 162, 162n12, 164, 174

Persuasion, 4, 6n3, 11, 12, 14, 21, 25n6, 70–75, 115–118, 137n2

Pivotal point, 25, 28, 33, 39, 45, 51, 83, 100, 116, 173

Pride and Prejudice, 4, 5, 9, 19, 42–50, 96–105

R

Rhythm, 4, 8–10, 19, 21, 29, 64, 119n8, 133, 136, 140, 152–154, 164, 174, 175

S

Sanditon, 75–76, 85, 118–130, 119n8, 137n2

Seasonal cyclicity, 21, 22, 28, 29, 56, 162

Seasons
- Autumn, 4, 20, 20n2, 23–25, 28, 29, 29n7, 31, 32, 36, 37, 39, 42, 44, 45, 47–50, 58, 61, 64, 69–71, 73–78, 139, 159, 162, 166, 173, 174
- Spring, 20, 20n2, 20n3, 23–25, 27–29, 33–36, 39, 41, 42, 45–48, 54–56, 63, 64, 67, 69, 71, 74, 76–78, 89, 101, 139, 173, 174
- Summer, 20, 20n2, 29, 35–37, 42, 46–49, 54–59, 63, 64, 68, 69, 71, 74, 76–78, 83, 114, 173
- Winter, 4, 20, 20n2, 25–27, 29, 32, 33, 39, 44, 45, 47, 58, 61, 63, 64, 67, 69, 73–78, 93, 162, 163, 173

Sense and Sensibility, 4, 36–42, 89–96

Space/time
- ballroom, 157, 160, 163, 164
- cyclical, 3–9, 23, 26, 34, 50, 62, 99, 108, 119, 148
- four dimensional, 3, 3n1
- linear, 2–9, 40, 52, 53, 56, 66, 74, 76, 87, 117, 130
- transitional, 6, 7, 34, 36, 70, 104, 113

Structure(s), 2, 10, 14, 19, 21, 31, 33, 36, 76–78, 157, 173–175

T

Trajectory, 7, 8, 22, 32, 69, 71, 104, 118, 142

W

Walking
- alone, 138, 140, 144–146, 148, 149, 152, 154
- emotional, 8, 13, 42, 101, 134, 136, 138–148, 150, 153, 154, 156, 165, 169, 174, 175
- physical, 8, 13, 133, 136–138, 143, 145–147, 150, 152, 154, 156, 169, 174, 175
- social, 21, 23, 34, 133–137, 141, 142, 145–147, 152, 154, 174, 175

Watsons, The, 75–76, 118–130, 162

Windows
- looking through, 129
- privacy, 128

Y

Year, 1, 4, 5, 20–23, 20n2, 20n3, 25–29, 29n7, 32, 33, 36, 37, 39, 40, 42, 43, 48n9, 49, 52–54, 57–61, 64, 69–71, 74–78, 85, 91, 92, 105, 106, 109, 111, 112, 114, 118, 121, 123, 141n3, 143, 145, 146, 157, 163–165, 173, 174